SpringerBriefs in Physics

SpringerBriefs in Physics are a series of slim high-quality publications encompassing the entire spectrum of physics. Manuscripts for SpringerBriefs in Physics will be evaluated by Springer and by members of the Editorial Board. Proposals and other communication should be sent to your Publishing Editors at Springer.

Featuring compact volumes of 50 to 125 pages (approximately 20,000–45,000 words), Briefs are shorter than a conventional book but longer than a journal article. Thus, Briefs serve as timely, concise tools for students, researchers, and professionals. Typical texts for publication might include:

- A snapshot review of the current state of a hot or emerging field
- A concise introduction to core concepts that students must understand in order to make independent contributions
- An extended research report giving more details and discussion than is possible in a conventional journal article
- A manual describing underlying principles and best practices for an experimental technique
- An essay exploring new ideas within physics, related philosophical issues, or broader topics such as science and society

Briefs allow authors to present their ideas and readers to absorb them with minimal time investment. Briefs will be published as part of Springer's eBook collection, with millions of users worldwide. In addition, they will be available, just like other books, for individual print and electronic purchase. Briefs are characterized by fast, global electronic dissemination, straightforward publishing agreements, easy-to-use manuscript preparation and formatting guidelines, and expedited production schedules. We aim for publication 8–12 weeks after acceptance.

More information about this series at http://www.springer.com/series/8902

Matthias Lienert · Sören Petrat · Roderich Tumulka

Multi-time Wave Functions

An Introduction

 Springer

Matthias Lienert
Fachbereich Mathematik
Eberhard-Karls-Universität Tübingen
Tübingen, Baden-Württemberg, Germany

Sören Petrat
Mathematical Sciences
Jacobs University Bremen
Bremen, Germany

Roderich Tumulka
Fachbereich Mathematik
Eberhard-Karls-Universität Tübingen
Tübingen, Baden-Württemberg, Germany

ISSN 2191-5423 ISSN 2191-5431 (electronic)
SpringerBriefs in Physics
ISBN 978-3-030-60690-9 ISBN 978-3-030-60691-6 (eBook)
https://doi.org/10.1007/978-3-030-60691-6

This Springer imprint is published by the registered company Springer Nature Switzerland AG
The registered company address is: Gewerbestrasse 11, 6330 Cham, Switzerland

Preface

A multi-time wave function is a quantum-mechanical wave function $\psi(x_1,\ldots,x_N)$ which, for N particles, depends on N space-time variables $x_i = (t_i, \mathbf{x}_i) \in \mathbb{R}^4$, and thus on N time variables. It provides a natural generalization of wave functions to relativistic space-time, in fact, a covariant position-representation of the quantum state. As we elucidate, the applicability of this simple and elegant concept is not limited to quantum mechanics but also comprises quantum field theory, where the configuration is allowed to consist of any (variable) number N of particles. Since simultaneous PDEs for several time variables can easily be inconsistent, the requirement of consistency restricts the possible time evolution laws for multi-time wave functions and leads us to certain types of laws that can be expressed in strikingly simple equations—as would seem very suitable for fundamental physical laws. So the concept of multi-time wave functions turns out to be rather deep, and a powerful tool in the study of relativistic quantum dynamics (and some other applications as well).

The concept was considered already in the early days of quantum mechanics, in fact, since 1929 by illustrious physicists such as Eddington [15], Gaunt [16], Mott [39], and, in particular, Dirac [8]. Surprisingly, the multi-time formalism was not studied comprehensively for many years. Much work on these wave functions, on the mathematically non-trivial consistency question of their evolution equations, and on ways of implementing interaction for them is rather recent. It is particularly this recent work that we aim at presenting here in the first book devoted to multi-time wave functions.

This book has originated from lectures at a spring school on multi-time wave functions that took place in Tübingen, Germany, on April 10–12, 2019. The lectures are intended for master-level students, doctoral students, and advanced undergraduates; in addition, they may also be helpful to researchers looking for an overview of the topic. As prerequisites, the lectures assume familiarity with basic notions of quantum mechanics and relativity. While in some places some supplementary material has been added, the chapters in this book retain the concise style of the lectures. We have included a number of exercises that were given to the participants at the spring school, and which may aid the reader's self-study. Each chapter in this book treats one of the lectures in the spring school, and the chapters were mainly written by the corresponding presenters, i.e., Chaps. 1 and 6 by Roderich Tumulka, Chaps. 2 and 4 by Sören Petrat, and Chaps. 3, 5, and 7 by Matthias Lienert.

We hope that this little volume will open a door for many a reader to an intriguing topic that is an active field of current research.

Tübingen, Germany   Matthias Lienert
August 2020   Sören Petrat
                                                                            Roderich Tumulka

Contents

Chapter 1
Introduction and Overview

In this chapter, we introduce the concept of a multi-time wave function and related basic notions and examples.

1.1 Relativistic Space-Time

We begin by introducing some notation. Let $\mathscr{M} = \mathbb{R}^4$ be Minkowski space-time with points $x = x^\mu = (x^0, \boldsymbol{x}) = (ct, \boldsymbol{x})$. Henceforth we set $c = 1$. The metric is

$$g_{\mu\nu} = \eta_{\mu\nu} = \begin{pmatrix} 1 & & & \\ & -1 & & \\ & & -1 & \\ & & & -1 \end{pmatrix}, \tag{1.1}$$

see Fig. 1.1. A 4-vector x^μ is called *causal* if it is timelike or lightlike (i.e., non-spacelike, or $x^\mu x_\mu \geq 0$). We follow the convention that the set future(x) contains also the future light cone of x and x itself. For a set $A \subseteq \mathscr{M}$, one writes

$$\text{future}(A) = \bigcup_{x \in A} \text{future}(x), \tag{1.2}$$

likewise for the past.

The *Lorentz group* $\mathscr{L} = O(1, 3)$ is the set of all Lorentz transformations including rotations in space, which form the rotation group $SO(3)$. The *restricted Lorentz group* or *proper Lorentz group* $\mathscr{L}^+ \subset \mathscr{L}$ contains those transformations that do not reverse either time or space. The symmetry group of Minkowski space-time is the *Poincaré group*

$$\mathscr{P} = \{x \mapsto a + \Lambda x \ : \ a \in \mathscr{M}, \Lambda \in \mathscr{L}\}, \tag{1.3}$$

which includes space-time translations; correspondingly, one defines $\mathscr{P}^+$.

© The Author(s), under exclusive license to Springer Nature Switzerland AG 2020
M. Lienert et al., *Multi-time Wave Functions*,
SpringerBriefs in Physics, https://doi.org/10.1007/978-3-030-60691-6_1

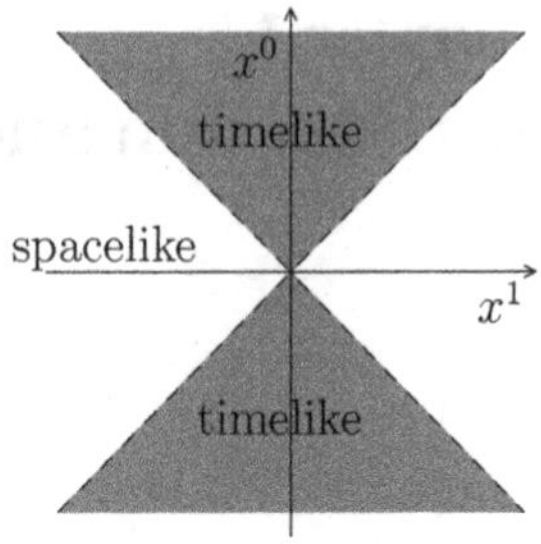

Fig. 1.1 Diagram of
Minkowski space-time in
1+1 dimensions. Vectors on
the light cone (dashed) are
called lightlike or null
vectors

Fig. 1.2 Three examples of
Cauchy surfaces

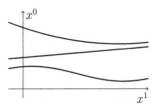

Definition 1.1 A *Cauchy surface* is a subset Σ of Minkowski space-time $\mathcal{M}$ which
is intersected by every inextensible causal (i.e., non-spacelike) curve exactly once.

A Cauchy surface is the kind of 3d surface on which we can expect that initial data
("Cauchy data") can be specified. We can imagine a Cauchy surface as a spacelike
surface that extends to infinity, see Fig. 1.2. (Technically, there are subtle differences,
but they will not be important for us.)

1.2 Dirac Equation

Here is a brief review of the Dirac equation.[1] For a wave function that takes values
$\psi(t, \boldsymbol{x}) \in \mathbb{C}^4$ in spin space $\mathbb{C}^4$, the Dirac equation reads (henceforth, $\hbar = 1$)

$$i\frac{\partial \psi}{\partial t} = -i\boldsymbol{\alpha} \cdot \nabla\psi + \beta m\psi =: H_{\text{Dirac}}\psi \qquad (1.4)$$

or, equivalently,

$$i\gamma^\mu \partial_\mu \psi = m\psi \quad \text{or} \quad i\partial\!\!\!/\psi = m\psi . \qquad (1.5)$$

Here, $\beta = \gamma^0$, $\alpha^i = \gamma^0\gamma^i$ ($i = 1, 2, 3$), and the "slash notation" means $v\!\!\!/ = v_\mu \gamma^\mu$ for
any Minkowski vector $v \in \mathcal{M}$. The gamma matrices satisfy the Clifford relation

$$\gamma^\mu\gamma^\nu + \gamma^\nu\gamma^\mu = 2\eta^{\mu\nu}\mathbb{1}, \qquad (1.6)$$

---

[1]Good introductory references are, for example, [46, 53].

in particular $\gamma^0\gamma^0 = 1$ (where 1 is the identity operator).

Hamiltonian formulation. The relevant Hilbert space is $\mathscr{H} = L^2(\mathbb{R}^3, \mathbb{C}^4)$ with inner product given by

$$\langle\phi|\psi\rangle = \int_{\mathbb{R}^3} d^3x\, \phi^\dagger(x)\,\psi(x) \tag{1.7}$$

with inner product in spin space

$$\phi^\dagger\psi = \sum_{s=1}^{4} \phi_s^*\,\psi_s\,. \tag{1.8}$$

H_{Dirac} is self-adjoint on its domain $H^1(\mathbb{R}^3, \mathbb{C}^4)$, the first Sobolev space. As a consequence, the time evolution operators $U_t := e^{-iHt} : \mathscr{H} \to \mathscr{H}$ are unitary.

Space-time formulation. We now regard the wave function as a function on space-time, $\psi : \mathscr{M} = \mathbb{R}^4 \to \mathbb{C}^4$. Spinors transform under $\Lambda \in \mathscr{L}^+$ according to

$$\pm S[\Lambda] : \mathbb{C}^4 \to \mathbb{C}^4\,, \tag{1.9}$$

where S is a (projective) *representation* of $\mathscr{L}^+$. Thus,

$$\psi'(x) = S[\Lambda]\,\psi(\Lambda^{-1}x)\,. \tag{1.10}$$

If Λ is a rotation through the angle φ, then $S(\Lambda)$ is a rotation in spin space through the angle $\varphi/2$; hence the name "spin-$\frac{1}{2}$."

Theorem 1.1 ([53]) *Every $\Lambda \in \mathscr{L}^+$ leaves γ^μ invariant, i.e.,*

$$\gamma^\mu = (\gamma')^\mu = S(\Lambda)\,\Lambda^\mu_\nu\gamma^\nu S(\Lambda)^{-1}\,. \tag{1.11}$$

Corollary 1.1 *The Dirac equation is Lorentz invariant.*

Remark. The inner product 1.8 is *not* Lorentz invariant. However, the following product is:

$$\overline{\phi}\psi := \phi^\dagger\gamma^0\psi\,. \tag{1.12}$$

Definition 1.2 For every $\psi : \mathscr{M} \to \mathbb{C}^4$, the *probability current 4-vector field* is

$$j^\mu(x) = \overline{\psi(x)}\gamma^\mu\psi(x)\,. \tag{1.13}$$

Here are some properties of the vector field j. First, it is defined in a covariant way. It is everywhere causal (i.e., timelike or lightlike),

$$j^\mu(x)\,j_\mu(x) \geq 0\,. \tag{1.14}$$

It is future-pointing, $j^0(x) \geq 0$. Its time component

$$j^0 = \overline{\psi}\gamma^0\psi = \psi^\dagger\gamma^0\gamma^0\psi = \psi^\dagger\psi = \sum_{s=1}^{4} |\psi_s|^2 = \rho \tag{1.15}$$

is the probability density according to Born's rule.

Exercise 1.1 (*continuity equation from Dirac equation*) Derive the continuity equation

$$\partial_\mu j^\mu = 0, \tag{1.16}$$

or, equivalently (writing $j = (\rho, \boldsymbol{j})$),

$$\partial_t \rho = -\mathrm{div}_3\, \boldsymbol{j}, \tag{1.17}$$

from the Dirac equation 1.5 and the definition 1.13 of j^μ.

Hint: Use that the adjoint of γ^μ is $\gamma^{\mu\dagger} = \gamma^0\gamma^\mu\gamma^0$, as can be verified in (e.g.) the standard representation

$$\gamma^0 = \begin{pmatrix} I & 0 \\ 0 & -I \end{pmatrix}, \quad \gamma^i = \begin{pmatrix} 0 & \sigma^i \\ -\sigma^i & 0 \end{pmatrix} \tag{1.18}$$

with $I = \left(\begin{smallmatrix} 1 & 0 \\ 0 & 1 \end{smallmatrix}\right)$ and Pauli matrices

$$\sigma^1 = \begin{pmatrix} 0 & 1 \\ 1 & 0 \end{pmatrix}, \quad \sigma^2 = \begin{pmatrix} 0 & -i \\ i & 0 \end{pmatrix}, \quad \sigma^3 = \begin{pmatrix} 1 & 0 \\ 0 & -1 \end{pmatrix}. \tag{1.19}$$

As will be shown in Exercise 1.4 in Sect. 1.5, it follows from 1.16 and the Ostrogradski–Gauss integral theorem (divergence theorem) in 4 dimensions that total probability is conserved in the sense that for any two Cauchy surfaces Σ, Σ',

$$\int_\Sigma d^3\sigma(x)\, j^\mu(x)\, n_\mu(x) = \int_{\Sigma'} d^3\sigma(x)\, j^\mu(x)\, n_\mu(x) \tag{1.20}$$

with

$$d^3\sigma(x) = d^3x\sqrt{\det(^3g(x))} \tag{1.21}$$

the 3-volume defined by the metric on Σ and $n_\mu(x)$ the future unit normal vector to Σ at $x \in \Sigma$, provided $j^\mu \to 0$ fast enough as "$x \to \infty$ spacelike."

Propagation locality. If $\psi(t = 0)$ is concentrated in $A \subset \mathbb{R}^3$ (i.e., $\psi(0, \boldsymbol{x}) = 0$ for all $\boldsymbol{x} \notin A$), then ψ is concentrated in future($\{0\} \times A$) $\cup$ past($\{0\} \times A$), the shaded region in Fig. 1.3. This statement means that the wave function cannot propagate faster than light.

Fig. 1.3 Propagation
locality means that a wave
function initially
concentrated in A is
concentrated in the shaded
region of space-time

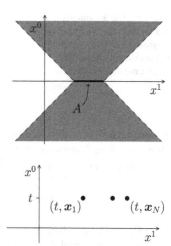

Fig. 1.4 Several space-time
points that are simultaneous
relative to the chosen
Lorentz frame

1.3 What is a Multi-time Wave Function?

The ordinary wave function of quantum mechanics of N particles

$$\varphi(t, \boldsymbol{x}_1, \ldots, \boldsymbol{x}_N) \tag{1.22}$$

evolves according to the Schrödinger equation

$$i\frac{\partial \varphi}{\partial t} = H\varphi \tag{1.23}$$

or, equivalently,

$$\varphi(t) = e^{-iHt}\varphi(0) \tag{1.24}$$

with H the Hamiltonian operator. It is uniquely determined by initial data $\varphi(0)$ on $\mathbb{R}^{3N}$. This description of the quantum state is not covariant, as it refers to space-time points $(t, \boldsymbol{x}_1), \ldots, (t, \boldsymbol{x}_N)$ that are simultaneous with respect to the chosen Lorentz frame, see Fig. 1.4.

The obvious alternative is to allow N space-time points as the arguments of the wave function. This results in a multi-time wave function (Fig. 1.5)

$$\psi(t_1, \boldsymbol{x}_1, \ldots, t_N, \boldsymbol{x}_N) = \psi(x_1, \ldots, x_N), \tag{1.25}$$

as suggested in particular by Dirac in 1932 [8] and already in 1929 by Eddington, Gaunt, and Mott [15, 16, 39]. (In contrast, in classical mechanics, multiple time variables are not as relevant [44].)

Fig. 1.5 The arguments of a
multi-time wave function are
N space-time points that do
not necessarily have the
same time coordinate

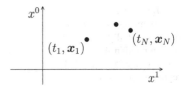

Example 1.1 (*non-interacting*) We begin with a case in which it is obvious how ψ should evolve: N non-interacting particles, for simplicity with $N = 2$. Let the wave function $\psi_{s_1 s_2}(x_1, x_2)$, $\psi : \mathscr{M}^2 \to \mathbb{C}^4 \otimes \mathbb{C}^4$, be given by

$$\psi(t_1, \cdot, t_2, \cdot) = e^{-i H_1 t_1 - i H_2 t_2} \psi(0, \cdot, 0, \cdot) \tag{1.26}$$

with $H_j = H_{\text{Dirac}}$ acting on $\boldsymbol{x}_j$, s_j (the free Hamiltonian for particle j); ψ is uniquely determined by initial data (Cauchy data) $\psi(0, 0) : \mathbb{R}^3 \times \mathbb{R}^3 \to \mathbb{C}^4 \otimes \mathbb{C}^4$ and obeys multi-time Schrödinger equations

$$i \frac{\partial \psi}{\partial t_1} = H_1 \psi \tag{1.27a}$$

$$i \frac{\partial \psi}{\partial t_2} = H_2 \psi. \tag{1.27b}$$

Note that the ordinary single-time wave function φ is contained in ψ according to

$$\varphi(t, \boldsymbol{x}_1, \boldsymbol{x}_2) := \psi(t, \boldsymbol{x}_1, t, \boldsymbol{x}_2) \tag{1.28}$$

and obeys, by the chain rule,

$$i \frac{\partial \varphi}{\partial t} = i \frac{\partial \psi}{\partial t_1}\bigg|_{(t,t)} + i \frac{\partial \psi}{\partial t_2}\bigg|_{(t,t)} = (H_1 + H_2)\varphi, \tag{1.29}$$

so $H = H_1 + H_2$ (non-interacting). The challenge is to include interaction. We will discuss how to do that in Chaps. 2–7.

Remarks.

- This construction also works for N particles.
- When we write the multi-time Eq. (1.27), not in Hamiltonian form, but in space-time form as in (1.5),

$$i \gamma_1^\mu \partial_{1,\mu} \psi = m \psi \tag{1.30a}$$

$$i \gamma_2^\mu \partial_{2,\mu} \psi = m \psi, \tag{1.30b}$$

where γ_j^μ is γ^μ acting on the index s_j (recall that ψ is a mapping from $\mathscr{M}^2$ to $\mathbb{C}^4 \otimes \mathbb{C}^4$) and $\partial_{j,\mu} = \partial/\partial x_j^\mu$ (also written $\partial_{x_j^\mu}$ in the following), we see that the system (1.30) is manifestly covariant and does not require a choice of a Lorentz frame (i.e., of a splitting of space-time into space and time).

- This construction of a non-interacting multi-time evolution works also in *curved* space-time. Here is a brief outline for the experts: For a single particle, the Dirac equation can also be considered for a curved space-time $(\mathcal{M}, g)$; then the wave function ψ is a cross-section of a vector bundle $\mathcal{D}$ over $\mathcal{M}$ consisting of 4-dimensional complex spin spaces; the gamma matrices become a cross-section of $\mathcal{D} \otimes \mathcal{D}^* \otimes T^*\mathcal{M}$ (with $T\mathcal{M}$ the tangent bundle); g and γ are used for defining a covariant derivative ∇_μ on $\mathcal{D}$; and the Dirac equation takes the form

$$i\gamma^\mu \nabla_\mu \psi = m\psi. \tag{1.31}$$

A multi-time wave function without interaction in $\mathcal{M}$ is then (say, for $N = 2$ particles) a cross-section of the vector bundle with fiber $\mathcal{D}_{x_1} \otimes \mathcal{D}_{x_2}$ at $(x_1, x_2) \in \mathcal{M} \times \mathcal{M}$ satisfying the multi-time equations

$$i\gamma_1^\mu \nabla_{1,\mu} \psi = m\psi \tag{1.32a}$$
$$i\gamma_2^\mu \nabla_{2,\mu} \psi = m\psi. \tag{1.32b}$$

- The non-interacting construction works also for the non-relativistic Schrödinger equation with free Hamiltonian $H_j = -\frac{1}{2m_j}\Delta_j$.

Example 1.2 (*second-order equations*) It is also possible to modify Example 1.1 for scalar wave functions $\psi : \mathcal{M}^2 \to \mathbb{C}$ and second-order Klein-Gordon equations,

$$\Box_1 \psi = m_1^2 \psi \tag{1.33a}$$
$$\Box_2 \psi = m_2^2 \psi \tag{1.33b}$$

with $\Box = \partial^\mu \partial_\mu = \partial_t^2 - \Delta$ the d'Alembertian operator.

Example 1.3 (*QFT*) Relativistic quantum theories are often formulated as quantum field theories (QFTs), so let us have a look at how they fit together with multi-time wave functions. Given a QFT in terms of field operators $\Phi(t, \boldsymbol{x}) = e^{iHt}\Phi(0, \boldsymbol{x})e^{-iHt}$ (in the Heisenberg picture) on Fock space $\mathcal{F}$, we can define a multi-time wave function ψ according to

$$\psi(x_1, \ldots, x_N) := \frac{1}{\sqrt{N!}} \langle \emptyset | \Phi(x_1) \cdots \Phi(x_N) | \Psi \rangle \tag{1.34}$$

with $|\emptyset\rangle$ the Fock vacuum and $|\Psi\rangle \in \mathcal{F}$ the state vector. Conversely, we will use ψ to construct QFTs in Chap. 4. The function ψ is a particle–position representation of the quantum state, expressed in a manifestly covariant way.

Example 1.4 (*detectors*) Multi-time wave functions also naturally come up when considering time measurements on several particles, specifically concerning the time

Fig. 1.6 Timelike Born rule
as in Example 1.4: Detectors
are placed along a timelike
3d surface Σ and register
when particles, initially in
the region Ω, arrive

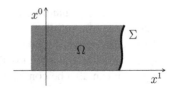

of detection as different particles may get detected at different times. We consider
ideal "hard" detectors that detect any particle as soon as it reaches the detector (as
opposed to "soft" detectors that may take a while to notice a particle in the detector
volume). Consider N non-interacting Dirac particles and hard detectors along the
timelike surface Σ that is the 3-dimensional boundary $\partial\Omega$ of the 4-dimensional region
Ω (see Fig. 1.6) except for the past boundary $\{x^0 = 0\}$. What is the probability of
detection at $(y_1, \ldots, y_N) \in \Sigma^N$? Put differently, what is the analog of the Born rule
for a timelike surface?

Answer [57, 58, 61]: Solve (1.27) for $1, \ldots, N$ on Ω^N with boundary condition
(BC)

$$\not{n}_j(x_j)\,\psi(x_1 \ldots x_N) = \not{u}_j(x_j)\,\psi(x_1 \ldots x_N) \tag{1.35}$$

for all $j \in \{1, \ldots, N\}, x_j \in \Sigma, x_1, \ldots, x_N \in \overline{\Omega}$ (where $\overline{\Omega}$ means the closure $\Omega \cup \partial\Omega$
of Ω). Here $n_\mu(x)$ is the outward unit normal vector to Σ at x, $u_\mu(x)$ is the unit timelike
vector of the local detector frame (tangent to Σ), and $\not{u}_j$ means $v_\mu\gamma^\mu$ acting on s_j.
Then

$$\mathrm{Prob}\Big(Y_1 \in \mathrm{d}^3 y_1, \ldots, Y_N \in \mathrm{d}^3 y_N\Big) =$$
$$\overline{\psi}(y_1 \ldots y_N)\,\not{n}_1(y_1) \cdots \not{n}_N(y_N)\,\psi(y_1 \ldots y_N)\,\mathrm{d}^3\sigma(y_1)\cdots\mathrm{d}^3\sigma(y_N). \tag{1.36}$$

Remark: An analogous rule can be formulated also for the non-relativistic case, using
the boundary condition

$$n(x_j) \cdot \nabla_j\psi(x_1 \ldots x_N) = i\kappa(x_j)\psi(x_1 \ldots x_N) \tag{1.37}$$

with detector-dependent $\kappa > 0$.

Example 1.5 (*scattering cross section*) Another situation involving different detection times for different particles (and thus another field of application for multi-time
wave functions) arises in scattering theory. To this end, consider "soft" detectors
along a distant sphere; that is, we consider $\Omega = \mathbb{R} \times B_R(\mathbf{0})$ in the limit $R \to \infty$,
where

$$B_r(\mathbf{y}) := \Big\{x \in \mathbb{R}^3 : |x - y| < r\Big\} \tag{1.38}$$

Fig. 1.7 Born rule for
curved Cauchy surfaces Σ

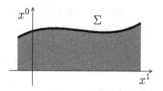

denotes the open 3-ball with radius r and center $\mathbf{y}$. Still, we use (1.27) and (1.36), but no boundary condition is necessary in the limit; no interaction is involved after a certain initial period because the particles then are far from each other.

Here, the answer is a probability distribution on $\big((\text{time axis}) \times \mathbb{S}^2\big)^N$ (with $\mathbb{S}^2 = \partial B_1(\mathbf{0})$ the unit sphere) given in the non-relativistic case by [13]

$$\lim_{R \to \infty} \text{Prob}\Big(Y_1 \in R \, dt_1 \, R \, d^2\omega_1, \dots, Y_N \in R \, dt_N \, R \, d^2\omega_N\Big) =$$

$$\Big|\mathscr{F}\varphi_0\Big(\tfrac{m\omega_1}{t_1}, \dots, \tfrac{m\omega_N}{t_N}\Big)\Big|^2 \, dt_1 \, d^2\omega_1 \cdots dt_N \, d^2\omega_N, \qquad (1.39)$$

where $\mathscr{F}$ means Fourier transformation and φ_0 the initial wave function after the interaction period.

Example 1.6 (*curved Born rule*) Even if we consider detectors placed along a *space-like* surface Σ (a Cauchy surface, see Fig. 1.7), the detection times of several particles will not be equal unless Σ is horizontal, and also here we are led to multi-time wave functions. So consider again N non-interacting Dirac particles. We write Y_k for the random point on Σ at which particle k gets detected. What is the joint probability distribution of $Y_1, \dots, Y_N$? Since $Y_1, \dots, Y_N$ do not necessarily all have the same time coordinate, we face again a problem of multiple time variables.

Answer (justified in Chap. 6): The probability distribution is again of the form (1.36), now with timelike normal vectors n_μ. In short, [4]

$$\text{``} \rho_\Sigma = |\psi_\Sigma|^2 \text{''} \qquad (1.40)$$

in the appropriate basis in spin space, and ψ_Σ, the wave function on Σ, is directly given by the multi-time wave function ψ according to

$$\psi_\Sigma(x_1, \dots, x_N) = \psi(x_1, \dots, x_N) \qquad (1.41)$$

for $x_1, \dots, x_N \in \Sigma$. In fact, $\psi_\Sigma \in \mathscr{H}_\Sigma$, which contains functions $\Sigma^N \to (\mathbb{C}^4)^{\otimes N}$ with inner product

$$\langle \chi | \varphi \rangle = \int_{\Sigma^N} d^3\sigma(x_1) \cdots d^3\sigma(x_N) \, \overline{\chi}(x_1 \dots x_N) \, \not{n}_1(x_1) \cdots \not{n}_N(x_N) \, \varphi(x_1 \dots x_N).$$

$$(1.42)$$

Conversely, a given family of functions ψ_Σ for every Cauchy surface Σ can be put together to a multi-time wave function ψ according to (1.41) if and only if

$$\psi_\Sigma(x_1, \ldots, x_N) = \psi_{\Sigma'}(x_1, \ldots, x_N) \tag{1.43}$$

whenever $x_1, \ldots, x_N \in \Sigma \cap \Sigma'$. We will see more details and the interacting case in Chap. 6.

1.4 Multi-time Schrödinger Equations

What kind of equations govern the time evolution of a multi-time wave function $\psi(x_1 \ldots x_N)$? What replaces the Schrödinger Eq. (1.23)? We want that $\psi(x_1 \ldots x_N)$ is determined by initial data (Cauchy data), specified for $t_1 = \cdots = t_N = 0$; this suggests [8]

$$i\frac{\partial \psi}{\partial t_1} = H_1 \psi \tag{1.44a}$$

$$\vdots$$

$$i\frac{\partial \psi}{\partial t_N} = H_N \psi, \tag{1.44b}$$

a form we have already seen in the non-interacting case (1.27). We may expect to be able to increase each time variable t_j individually by means of the j-th equation in (1.44). (An alternative to the form (1.44) would be to use integral equations, see Chap. 7.)

We call H_j the partial Hamiltonian for particle j. Since we expect as a general relation between the single-time wave function φ and the multi-time wave function ψ that

$$\varphi(t, x_1, \ldots, x_N) = \psi(t, x_1, \ldots, t, x_N), \tag{1.45}$$

it follows again that

$$H = \sum_{j=1}^{N} H_j \bigg|_{(t,t,\ldots,t)}, \tag{1.46}$$

while we do not require that H (or H_j) is a free Hamiltonian.

Consistency question. A system of equations such as (1.44) does not automatically possess solutions. In fact, its solvability requires conditions on the H_j called consistency conditions [4]. Here is why. Suppose for simplicity that each $H_j : L^2(\mathbb{R}^{3N}, \mathbb{C}) \to L^2(\mathbb{R}^{3N}, \mathbb{C})$ is time independent, and consider only $N = 2$ particles. Think of ψ as a function of t_1, t_2 with values in the Hilbert space $L^2(\mathbb{R}^6, \mathbb{C})$. If ψ solves both equations of (1.44), then

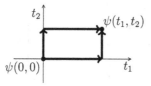

Fig. 1.8 The plane spanned by two time axes, and two paths in it corresponding to the two ways of obtaining $\psi(t_1, t_2)$ from $\psi(0, 0)$ as in (1.47)

$$e^{-i H_2 t_2} e^{-i H_1 t_1} \psi(0, 0) = \psi(t_1, t_2) = e^{-i H_1 t_1} e^{-i H_2 t_2} \psi(0, 0), \qquad (1.47)$$

see Fig. 1.8.

If the initial datum $\psi(0, 0)$ can be arbitrary, this requires that

$$\left[e^{-i H_1 t_1}, e^{-i H_2 t_2} \right] = 0 \ \forall t_1, t_2 \qquad (1.48)$$

or, equivalently,

$$\left[H_1, H_2 \right] = 0, \qquad (1.49)$$

the consistency condition. The more general form of the consistency condition for time-dependent Hamiltonians reads

$$\left[i \frac{\partial}{\partial t_1} - H_1, i \frac{\partial}{\partial t_2} - H_2 \right] = 0. \qquad (1.50)$$

The consistency question will be discussed more deeply in Chap. 2.

Example 1.7 (*quantum control*) [18, 48] A problem mathematically equivalent to the consistency of multi-time equations arises in quantum control theory. Let $t_1 = t$ be physical time and $t_2, \ldots, t_N$ some parameters that experimenters can control (external fields). We vary $t_2(t), \ldots, t_N(t)$ for $t \in [0, T]$ from $t_j(0)$ to $t_j(T)$. If the equations satisfy the consistency condition, then $\psi(T)$ depends only on the final parameters $t_j(T)$ but not on the path $t_j(t)$ in parameter space.

Definition 1.3 The set of *spacelike configurations* of N particles is

$$\mathscr{S}_N := \left\{ (x_1, \ldots, x_N) \in \mathscr{M}^N \ : \ \forall j, k : (x_j - x_k)^\mu (x_j - x_k)_\mu < 0 \text{ or } x_j = x_k \right\}, \qquad (1.51)$$

see Fig. 1.9. Often, the multi-time wave function is defined *only* on the spacelike configurations, i.e., $\psi : \mathscr{S}_N \to \mathbb{C}^K$ instead of $\psi : \mathscr{M}^N \to \mathbb{C}^K$. It then follows that in the non-relativistic limit $c \to \infty$ (see Fig. 1.10), the domain $\mathscr{S}_N$ of ψ approaches the set of simultaneous configurations (i.e., with equal time coordinates as in Fig. 1.4),

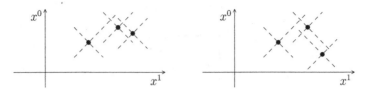

Fig. 1.9 LEFT: A spacelike configuration (light cones dashed). RIGHT: A non-spacelike config-uration; some pairs of events are spacelike separated, but not all

Fig. 1.10 If c is large, then light cones are wide

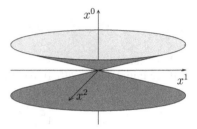

which is why multi-time wave functions usually do not arise in the non-relativistic context.[2]

Tomonaga–Schwinger approach. [47, 54] This is closely related to multi-time wave functions. Suppose we have

- a Hilbert space $\mathcal{H}_\Sigma$ for every Cauchy surface Σ,
- unitary time evolution operators $U_\Sigma^{\Sigma'} : \mathcal{H}_\Sigma \to \mathcal{H}_{\Sigma'}$,
- a wave function $\psi_{\Sigma'} = U_\Sigma^{\Sigma'} \psi_\Sigma$ in $\mathcal{H}_\Sigma$ for every Σ
- and unitary operators $F_\Sigma^{\Sigma'} : \mathcal{H}_\Sigma \to \mathcal{H}_{\Sigma'}$ representing the *free* time evolution.

Fix any Cauchy surface Σ_0 as the "initial surface." Tomonaga and Schwinger used the *interaction picture*, in which one keeps track only of the deviation from the free evolution and sets

$$\widetilde{\psi}_\Sigma := F_\Sigma^{\Sigma_0} U_{\Sigma_0}^\Sigma \psi_{\Sigma_0} . \tag{1.52}$$

Then all $\widetilde{\psi}_\Sigma$ lie in the same Hilbert space $\mathcal{H}_{\Sigma_0}$ and evolve according to the *Tomonaga–Schwinger equation*: For Σ' infinitesimally close to Σ as in Fig. 1.11,

$$i(\widetilde{\psi}_{\Sigma'} - \widetilde{\psi}_\Sigma) = \int_\Sigma^{\Sigma'} \mathrm{d}^4 x \, \mathcal{H}_I(x) \, \widetilde{\psi}_\Sigma . \tag{1.53}$$

---

[2]Sometimes, the concept of multi-time wave functions gets mixed up with the idea that space-time might have more than 4 dimensions, of which more than 1 might be timelike. Actually, the latter idea is physically much less plausible and has little if any similarity with multi-time wave functions, as discussed in [33].

Fig. 1.11 Two Cauchy
surfaces that are
infinitesimally close

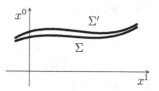

The operator-valued function $\mathcal{H}_I(x)$ is called the interaction Hamiltonian density. Also here, the Hamiltonian operators need to satisfy a consistency condition for the evolution to exist; it reads

$$\left[\mathcal{H}_I(x), \mathcal{H}_I(y)\right] = 0 \quad \text{for spacelike separated } x, y. \tag{1.54}$$

More will be said about this in Sect. 4.6.

Quantum theories without observers. Standard quantum theory is notoriously vague when we want to analyze the measurement process and when we ask what happens in reality [19]. Proposals for a firm foundation of quantum mechanics, in which observers obey the same laws as any other system of electrons and nuclei, include Bohmian mechanics [14, 20] and collapse theories [17]. What kind of wave functions do they use?

Bohmian mechanics uses the wave function only on special Cauchy surfaces, those belonging to a certain foliation (slicing) of space-time [12, 56]; as a consequence, it does not require multi-time wave functions for its definition. However, if that slicing consists, not of parallel hyperplanes, but more general, curved Cauchy surfaces Σ, then it is convenient to define ψ_Σ from a multi-time wave function ψ via (1.41) [12]. Moreover, relevant facts about Bohmian mechanics, such as the fact that observers cannot determine the preferred foliation from their observations [12, 36], are best proved using the fact that the wave function could be extended to arbitrary Cauchy surfaces (and thus to multi-time wave functions); actually, Theorem 6.1 of Chap. 6 plays a key role for this particular fact [36].

Let us turn to collapse theories [17] such as the Ghirardi-Rimini-Weber (GRW) theory. Relativistic collapse theories are naturally formulated in terms of wave functions ψ_Σ associated with Cauchy surfaces Σ [1, 2, 55, 59]. Their evolution is a combination of a unitary evolution $U_\Sigma^{\Sigma'}$ as in the Tomonaga-Schwinger approach and a collapse process that will push ψ_Σ towards macroscopically definite states in a stochastic fashion. As a consequence, these ψ_Σ cannot necessarily be put together to a multi-time wave function ψ according to (1.41) because the same spacelike configuration $(x_1, \ldots, x_N)$ may lie on several Cauchy surfaces Σ, Σ' for which $\psi_\Sigma(x_1 \ldots x_N) \neq \psi_{\Sigma'}(x_1 \ldots x_N)$ because a collapse occurred between Σ and Σ'. However, the study of the unitary part of the evolution is closely linked to multi-time wave functions.

1.5   Exercises

Exercise 1.2 (*continuity equation for N free particles*) Suppose that $\psi : \mathbb{R}^{4N} \to (\mathbb{C}^4)^{\otimes N}$ satisfies the free multi-time Dirac equations $i\gamma_j^\mu \partial_{x_j^\mu} \psi = m\psi$, where γ_j^μ is γ^μ acting on s_j. Let $\overline{\psi} = \psi^\dagger \gamma_1^0 \dots \gamma_N^0$ and

$$j^{\mu_1 \cdots \mu_N}(x_1 \dots x_N) = \overline{\psi}(x_1 \dots x_N) \gamma_1^{\mu_1} \cdots \gamma_N^{\mu_N} \psi(x_1 \dots x_N). \qquad (1.55)$$

Show that

$$\partial_{x_j^{\mu_j}} j^{\mu_1 \cdots \mu_N}(x_1 \dots x_N) = 0 \qquad (1.56)$$

for all $j = 1, \dots, N$.

Exercise 1.3 (*spacelike configurations*) Consider the case of $N = 2$ particles. We denote the set of spacelike configurations (including collision configurations) by

$$\mathscr{S} = \{(x_1, x_2) \in \mathbb{R}^4 \times \mathbb{R}^4 : |x_1^0 - x_2^0| < |\boldsymbol{x}_1 - \boldsymbol{x}_2| \text{ or } x_1^0 = x_2^0, \boldsymbol{x}_1 = \boldsymbol{x}_2\}. \quad (1.57)$$

Show that $\mathscr{S}$ is the smallest Poincaré invariant set which contains the equal-time configurations

$$\mathscr{E} = \{(x_1, x_2) \in \mathbb{R}^4 \times \mathbb{R}^4 : x_1^0 = x_2^0\}. \qquad (1.58)$$

Exercise 1.4 (*probability conservation on Cauchy surfaces*) Let $N \in \mathbb{N}$ and $\psi \in C^1(\mathbb{R}^{4N}, \mathbb{C}^{4^N})$ (the space of continuously differentiable functions from $\mathbb{R}^{4N}$ to $\mathbb{C}^{4^N}$) be a solution of the free multi-time Dirac equations $(i\gamma_k^\mu \partial_{x_k^\mu} - m_k)\psi = 0$, $k = 1, \dots, N$ which is compactly supported in space for all fixed time variables. For every smooth Cauchy surface Σ with future-pointing unit normal vector field n, we define

$$P(\Sigma) = \int_\Sigma d\sigma(x_1) \cdots \int_\Sigma d\sigma(x_N) \, \overline{\psi}(x_1, \dots, x_N) \, \not{n}_1(x_1) \cdots \not{n}_N(x_N) \, \psi(x_1, \dots, x_N).$$

$$(1.59)$$

(a) Show that $P(\Sigma) = P(\Sigma')$ for all pairs of smooth Cauchy surfaces Σ, Σ'.
   Hint: Apply the Ostrogradski–Gauss integral theorem to the volume between Σ and Σ', with a limit of mantle surfaces moving to spacelike infinity.
(b) Let ψ, ϕ be two solutions of the same initial value problem $\psi|_{\Sigma_0^N} = \phi|_{\Sigma_0^N} = \psi_0$ for some given function $\psi_0 \in C_c^\infty(\Sigma_0^N, \mathbb{C}^{4^N})$ (i.e., a smooth function with compact support). Show that (a) implies $\psi|_{\Sigma^N} = \phi|_{\Sigma^N}$ for all smooth Cauchy surfaces Σ.

Fig. 1.12 Σ_0 and Σ_t are parts of horizontal (equal time) surfaces, Σ^s is part of the past light cone of y. Σ_0, Σ_t and Σ^s enclose a volume in $\mathbb{R}^4$, a truncated cone

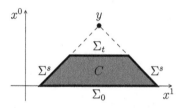

Hint: You can use that

$$\overline{\psi}(x_1, \ldots, x_N)\, \slashed{n}_1(x_1) \cdots \slashed{n}_N(x_N)\, \psi(x_1, \ldots, x_N) \geq 0 \qquad (1.60)$$

for all future-pointing timelike or lightlike vector fields n.

Exercise 1.5 (*finite propagation speed and domain of dependence*)

(a) Consider the 4-volume C depicted in a 2-dimensional way in Fig. 1.12. C is the volume enclosed by Σ_0, Σ_t, and Σ^s. Let $j : \mathbb{R}^4 \to \mathbb{R}^4$ be a continuously differentiable vector field. Taking $\mathbb{R}^4$ as a coordinate space with Euclidean metric, what are the outward unit normal vectors for Σ_0, Σ_t, and Σ^s? Then, write out explicitly the 4-dimensional Ostrogradski–Gauss integral theorem for $\int_C \mathrm{d}^4x \, \mathrm{div}_4(j)$.

(b) Consider the one-particle Dirac equation $i\gamma^\mu\partial_\mu\psi = \big(m + V(x)\big)\psi$ with smooth self-adjoint external potential $V \in C^\infty(\mathbb{R}^3, \mathbb{C}^{4\times4})$. For smooth initial data $\psi_0 \in C^\infty(\mathbb{R}^3, \mathbb{C}^4)$ it is known that there is a unique smooth solution $\psi \in C^\infty(\mathbb{R}^4, \mathbb{C}^4)$. Prove that $\psi(t, x)$ for $x \in B_{T-t}(y)$ is uniquely determined by specifying the initial conditions on $B_T(y)$.
Hint: Because of linearity, it suffices to consider $\psi(0, x) = 0$. Use $\partial_\mu j^\mu = 0$ and part (a).

Chapter 2
Consistency Conditions and Interaction Potentials

Most of the rest of this book is devoted to the question of how interaction can be introduced in the multi-time framework in a consistent way. In Sect. 1.4 we already saw that for two particles we require that $[H_1, H_2] = 0$ in order to define the multi-time wave function

$$e^{-i H_2 t_2} e^{-i H_1 t_1} \psi(0, 0) = \psi(t_1, t_2) = e^{-i H_1 t_1} e^{-i H_2 t_2} \psi(0, 0). \qquad (2.1)$$

For non-interacting particles as in Example 1.1, the condition $[H_1, H_2] = 0$ is certainly satisfied, since H_1 and H_2 act on different particles (i.e., $H_1 = H_{\text{Dirac}} \otimes \mathbb{1}$ and $H_2 = \mathbb{1} \otimes H_{\text{Dirac}}$). In general, a common solution of the N multi-time equations (1.44) will exist (for all initial conditions) only for *some* H_j, namely those which satisfy a *consistency condition* such as (1.50). In this chapter, we show that the consistency condition is indeed necessary and sufficient for the N multi-time equations (1.44). Furthermore, we explore the possibility of including interaction potentials into H_j. The conclusion will be that this is generally not consistent, except when the interaction has very short range and we consider the multi-time wave function only on a set smaller than the spacelike configurations (and not Lorentz invariant).

2.1 Rigorous Formulations of Consistency Conditions

First, we note that consistency conditions arise as necessary conditions for the existence of solutions to the multi-time equations (1.44). Let us consider a solution $\psi : \mathbb{R}^{4N} \to \mathbb{C}^K$ (with $\mathbb{C}^K$ the spin space) in $C^2(\mathbb{R}^{4N}, \mathbb{C}^K)$ (i.e., twice continuously differentiable). Then applying $i \frac{\partial}{\partial t_j} - H_j$ to ψ gives zero, which yields in particular that

$$\left[i \frac{\partial}{\partial t_j} - H_j, i \frac{\partial}{\partial t_k} - H_k \right] \psi = 0 \qquad (2.2)$$

© The Author(s), under exclusive license to Springer Nature Switzerland AG 2020
M. Lienert et al., *Multi-time Wave Functions*,
SpringerBriefs in Physics, https://doi.org/10.1007/978-3-030-60691-6_2

for all $j \neq k$ from $\{1, \ldots, N\}$. Now any given $t_1, \ldots, t_N$ can be times for specifying the initial data. If the initial datum is arbitrary, the left-hand side of (2.2) must vanish for every ψ, so the commutator itself must vanish,

$$\left[i\frac{\partial}{\partial t_j} - H_j, i\frac{\partial}{\partial t_k} - H_k \right] = 0 . \tag{2.3}$$

Note that (2.2) can be rewritten as

$$\underbrace{\left[i\frac{\partial}{\partial t_j}, i\frac{\partial}{\partial t_k} \right]\psi}_{=0\ (\psi \in C^2)} - \underbrace{\left[i\frac{\partial}{\partial t_j}, H_k \right]\psi}_{=i\left(\frac{\partial H_k}{\partial t_j}\right)\psi} - \underbrace{\left[H_j, i\frac{\partial}{\partial t_k} \right]\psi}_{=-i\left(\frac{\partial H_j}{\partial t_k}\right)\psi} + \left[H_j, H_k \right]\psi = 0 , \tag{2.4}$$

so (2.3) means

$$i\frac{\partial H_j}{\partial t_k} - i\frac{\partial H_k}{\partial t_j} + [H_j, H_k] = 0. \tag{2.5}$$

For time-independent H_j, this reduces to the condition $[H_j, H_k] = 0$ that we already derived in Chap. 1. The need for consistency conditions was made explicit early on by Bloch [4] and studied for more general multi-time PDEs by Ghiu and Udrişte [18], among others.

For the following discussion, it will be advantageous to consider a Hilbert space framework in which the multi-time wave function is of the form $\psi : \mathbb{R}^N \to \mathcal{H}$ with $\mathbb{R}^N$ the space spanned by the N time axes and $\mathcal{H}$ a Hilbert space such as $\mathcal{H} = L^2(\mathbb{R}^{3N}, \mathbb{C}^K)$ whose elements represent the dependence on the spatial variables. On the one hand, this framework means that we choose a fixed Lorentz frame and treat space and time variables differently; this does not cause problems, as it is just a method of proof. On the other hand, it means that we take ψ, when written as a $\mathbb{C}^K$-valued function of $(t_1, \boldsymbol{x}_1, \ldots, t_N, \boldsymbol{x}_N)$, to be defined also for non-spacelike configurations. This is an easier case that we focus on first; later, in Theorems 2.3, 2.4 and Assertion 4.1, we will also cover the case in which ψ is defined only for spacelike configurations.

Theorem 2.1 ([42]) *Let $H_1, \ldots, H_N$ be self-adjoint operators on $\mathcal{H}$. Then a solution to the multi-time equations (1.44) exists for all initial data $\psi(0, \ldots, 0) \in \mathcal{H}$ if and only if $[H_j, H_k] = 0$ for all $j \neq k$.*

With "solution" we here mean that

$$\psi(t_1, \ldots, t_N) = e^{-iH_j t_j} \psi(t_1, \ldots, t_{j-1}, 0, t_{j+1}, \ldots, t_N) \tag{2.6}$$

for all $(t_1, \ldots, t_N) \in \mathbb{R}^N$. For unbounded H_j, the condition "$[H_j, H_k] = 0$" means that the corresponding spectral projections commute, i.e.,

$$[\mathbb{1}_{A_j}(H_j), \mathbb{1}_{A_k}(H_k)] = 0 \tag{2.7}$$

for all $A_j, A_k \subseteq \mathbb{R}$.

Proof As already noted before, existence for all initial data requires that $[e^{-iH_j t_j}, e^{-iH_k t_k}] = 0$, which is equivalent to $[H_j, H_k] = 0$. ☐

Next, we consider time-dependent $H_j = H_j(t_1, \ldots, t_N)$. Let $\mathscr{L}(\mathscr{H})$ denote the set of bounded operators $\mathscr{H} \to \mathscr{H}$. For simplicity, let us state and prove the following theorem for H_j bounded and such that all $H_j(t_1, \ldots, t_N)$ are smooth as functions from $\mathbb{R}^N$ into $\mathscr{L}(\mathscr{H})$.

Theorem 2.2 ([42]) *Let $H_j : \mathbb{R}^N \to \mathscr{L}(\mathscr{H})$ be smooth for all $j = 1, \ldots, N$. Then a solution to the multi-time equations (1.44) exists for all initial data $\psi(0, \ldots, 0) \in \mathscr{H}$ if and only if*

$$\left[i\frac{\partial}{\partial t_j} - H_j, i\frac{\partial}{\partial t_k} - H_k \right] = 0 \tag{2.8}$$

for all $j \neq k$.

Proof We use the Dyson series. For a single time variable, the Dyson series is used in the case of a time-dependent Hamiltonian $H(t)$. It is an expression for the unitary time evolution operator $U(t, s)$ from time s to time t. Here, for the single-time wave function φ evolving according to

$$i\frac{d\varphi}{dt} = H(t)\varphi(t), \tag{2.9}$$

the defining property of $U(t, s)$ is that

$$U(t, s)\varphi(s) = \varphi(t). \tag{2.10}$$

While in the case of time-independent Hamiltonian H,

$$U(t, s) = e^{-iH(t-s)} = \mathbb{1} + \sum_{n=1}^{\infty} \frac{(-iH(t-s))^n}{n!} \tag{2.11a}$$

$$= \mathbb{1} + \sum_{n=1}^{\infty} (-i)^n \underbrace{\int_s^t dT_1 \int_s^{T_1} dT_2 \cdots \int_s^{T_{n-1}} dT_n}_{=(t-s)^n/n!} H^n, \tag{2.11b}$$

the corresponding expression in the case of time-dependent $H(t)$, the Dyson series, reads

$$U(t,s) = \mathbb{1} + \sum_{n=1}^{\infty} (-i)^n \int_s^t dT_1 \int_s^{T_1} dT_2 \cdots \int_s^{T_{n-1}} dT_n \, H(T_1)H(T_2)\cdots H(T_n) \quad \text{(2.12a)}$$

$$=: \mathscr{T} e^{-i\int_s^t H(T)\, dT} \quad \text{(2.12b)}$$

with $\mathscr{T}$ the "time ordering" of any sum of products of $H(t)$. Note that $U(t,s)U(s,t) = \mathbb{1}$ and $U(t,s)U(s,r) = U(t,r)$.

Now we apply this to the multi-time wave function ψ. For simplicity of notation, we take $N = 2$; then

$$\psi(t_1, t_2) = U(t_1, s_1; t_2)\psi(s_1, t_2) \quad \text{(2.13)}$$

with

$$U(t_1, s_1; t_2) = \mathbb{1} + \sum_{n=1}^{\infty} (-i)^n \int_{s_1}^{t_1} dT_1 \int_{s_1}^{T_1} dT_2 \cdots \int_{s_1}^{T_{n-1}} dT_n \, H_1(T_1, t_2)\cdots H_1(T_n, t_2).$$
$$\text{(2.14)}$$

Similarly to the time-independent case, we can then again write $\psi(t_1, t_2)$ in two ways,

$$\psi(t_1, t_2) = U(t_1, s_1; t_2)\, U(s_1; t_2, s_2)\, \psi(s_1, s_2) \quad \text{(2.15a)}$$
$$= U(t_1; t_2, s_2)\, U(t_1, s_1; s_2)\, \psi(s_1, s_2). \quad \text{(2.15b)}$$

So, the existence of a solution for every $\psi(s_1, s_2)$ (or for every $\psi(0,0)$) is equivalent to

$$U(t_1, s_1; t_2)\, U(s_1; t_2, s_2) = U(t_1; t_2, s_2)\, U(t_1, s_1; s_2) \quad \text{(2.16)}$$

for all $s_1, s_2, t_1, t_2 \in \mathbb{R}$. Put differently, the evolution U_γ along the closed path γ shown in Fig. 2.1 is the identity.

It will now be convenient to rename $s_k \to t_k$, $t_k \to t_k + \Delta t_k$. For simplicity, let $\Delta t_1 = \Delta t_2 = \Delta t$. Then for any $s_k \in [t_k, t_k + \Delta t]$, the Taylor expansion to first order of $H_j(s_1, s_2)$ around (t_1, t_2) reads

$$H_j(s_1, s_2) = H_j(t_1, t_2) + \sum_{k=1}^{2} (s_k - t_k)\frac{\partial H_j}{\partial t_k}(t_1, t_2) + o(\Delta t), \quad \text{(2.17)}$$

Fig. 2.1 Closed path in the plane spanned by two time axes

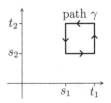

where the "little o" notation means a term such that

$$\frac{o(\Delta t)}{\Delta t} \xrightarrow{\Delta t \to 0} 0 . \tag{2.18}$$

Hence,

$$U(t_1 + \Delta t, t_1; t_2)$$

$$= \mathbb{1} - i \int_{t_1}^{t_1+\Delta t} dT_1 \, H_1(T_1, t_2) - \int_{t_1}^{t_1+\Delta t} dT_1 \int_{t_1}^{T_1} dT_2 \, H_1(T_1, t_2) \, H_1(T_2, t_2) + \mathcal{O}(\Delta t^3) \tag{2.19a}$$

$$= \mathbb{1} - i \, H_1(t_1, t_2) \, \Delta t - \frac{i}{2} \frac{\partial H_1}{\partial t_1}(t_1, t_2) \, \Delta t^2 + \frac{1}{2} H_1(t_1, t_2)^2 \Delta t^2 + o(\Delta t^2) , \tag{2.19b}$$

where the "big $\mathcal{O}$" means a term such that $\mathcal{O}(\Delta t^3)/\Delta t^3$ stays bounded as $\Delta t \to 0$. A computation similar to (2.19) for the other line segments in γ yields that (2.16) is equivalent to

$$0 = \left(-[H_1, H_2] - i \frac{\partial H_1}{\partial t_2} + i \frac{\partial H_2}{\partial t_1} \right) \Bigg|_{(t_1, t_2)} \Delta t^2 + o(\Delta t^3) . \tag{2.20}$$

The concatenation of many loops around little squares to provide the loop around a big rectangle then proves the statement. □

One can also approach the consistency condition and the proof from a different point of view. Define, for *any* path γ in the $t_1 t_2$-plane (see Fig. 2.2), the evolution operator

$$U_\gamma := \mathcal{T} e^{-i \int_\gamma \sum_j H_j \, dt_j} , \tag{2.21}$$

where $\mathcal{T}$ means "path ordering" relative to γ (i.e., that in products of Hamiltonians, those associated with earlier points on the path get applied first).

Then consistency is equivalent to path-independence of U_γ, i.e., to the condition that $U_\gamma = U_{\gamma'}$ whenever two paths γ, γ' have the same start point and the same end point. In differential geometry language, ψ is a cross-section of a vector bundle over the base manifold $\mathbb{R}^2$ with fiber space $\mathcal{H}$ (at each base point); for the covariant

Fig. 2.2 Arbitrary path in the $t_1 t_2$-plane

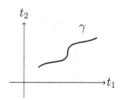

derivative defined by $\nabla_j := \partial_j - i H_j$, U_γ represents the parallel transport along γ. Path independence of U_γ is equivalent to the condition that the curvature F_{jk} of the covariant derivative vanishes. In fact, the curvature is given by

$$F_{jk} = i \frac{\partial H_j}{\partial t_k} - i \frac{\partial H_k}{\partial t_j} + [H_j, H_k], \tag{2.22}$$

so we obtain again the same consistency condition (2.5).

Consistency condition for higher-order equations. Also for second-order PDEs as in Example 1.2, and for higher order PDEs, one can formulate consistency conditions. They have been worked out by Nickel [41, Chap. 5] for the system of nth order equations

$$\partial_{t_1}^n \psi = L_1 \psi \tag{2.23a}$$

$$\vdots$$

$$\partial_{t_N}^n \psi = L_N \psi \tag{2.23b}$$

with unknown function $\psi : (\mathbb{R}^4)^N \to \mathbb{C}^K$ and linear differential operators L_j that involve only spatial derivatives (of arbitrary order) but no time derivatives. As initial conditions (Cauchy data) at $t_1 = \cdots = t_N = 0$, we can specify functions of $\boldsymbol{x}_1, \ldots, \boldsymbol{x}_N$ for ψ, $\partial_{t_j} \psi$, $\partial_{t_j} \partial_{t_k} \psi$, and generally $\partial_{t_1}^{\alpha_1} \cdots \partial_{t_N}^{\alpha_N} \psi$ for all $\alpha = (\alpha_1, \ldots, \alpha_N) \in \{0, 1, \ldots, n-1\}^N$. In particular, we need n^N initial values at every $(\boldsymbol{x}_1, \ldots, \boldsymbol{x}_N)$.

As shown by Nickel [41], the following conditions are necessary for the consistency of (2.23) for $n > 1$:

$$\frac{\partial L_k}{\partial t_j} = 0 \quad \forall j \neq k \tag{2.24a}$$

$$[L_j, L_k] = 0 \quad \forall j \neq k. \tag{2.24b}$$

It may be surprising that, while for equations of order $n = 1$, only a combination of the time derivatives and commutators has to vanish, for $n > 1$ they need to vanish individually. (In fact, this need not be the case if the L_j can contain time derivative operators.) To understand how the conditions (2.24) arise, we consider the case of $n = 2$ and $N = 2$, in which we can specify ψ, $\partial_{t_1} \psi$, $\partial_{t_2} \psi$, and $\partial_{t_1} \partial_{t_2} \psi$ as initial data. Since $\partial_{t_1}^2 \psi = L_1 \psi$ and $\partial_{t_2}^2 \psi = L_2 \psi$, we know that

$$\partial_{t_1}^2 \partial_{t_2}^2 \psi = \partial_{t_1}^2 (L_2 \psi) = (\partial_{t_1}^2 L_2)\psi + 2(\partial_{t_1} L_2)(\partial_{t_1} \psi) + L_2 L_1 \psi. \tag{2.25}$$

On the other hand, for ψ that is four times continuously differentiable,

$$\partial_{t_1}^2 \partial_{t_2}^2 \psi = \partial_{t_2}^2 \partial_{t_1}^2 \psi = (\partial_{t_2}^2 L_1)\psi + 2(\partial_{t_2} L_1)(\partial_{t_2} \psi) + L_1 L_2 \psi. \tag{2.26}$$

Since $\partial_{t_1}\psi$ and $\partial_{t_2}\psi$ can be chosen independently of ψ and of each other, (2.25) and (2.26) can only be equal for all initial data if the terms involving $\partial_{t_1}\psi$ and $\partial_{t_2}\psi$ vanish individually. Thus, $\partial_{t_1}L_2 = 0 = \partial_{t_2}L_1$, which automatically implies that $\partial_{t_1}^2 L_2 = 0 = \partial_{t_2}^2 L_1$ and thus leaves us with the further condition that $L_2L_1 = L_1L_2$.

2.2   Interaction Potentials

In Sect. 1.4 we saw that the single-time wave function $\varphi(t, x_1, \ldots, x_N) = \psi((t, x_1),$ $\ldots, (t, x_N))$ satisfies

$$i\frac{\partial\varphi}{\partial t} = \underbrace{\sum_{j=1}^{N} H_j\Big|_{(t,\ldots,t)}}_{H(t)} \varphi. \tag{2.27}$$

Therefore, the Hamiltonian H might provide some suggestion for what the partial Hamiltonians H_j should be. In non-relativistic quantum mechanics, one often considers H of the form

$$H = \underbrace{\sum_{i=1}^{N}\left(H_i^{\text{free}} + W(t, x_i)\right)}_{\text{non-interacting part}} + \underbrace{\sum_{1\leq i<j\leq N} V(x_i - x_j)}_{\text{interaction}}, \tag{2.28}$$

for example with the Coulomb potential

$$V(x) = \frac{\lambda}{|x|} \tag{2.29}$$

and $H_i^{\text{free}} = -\Delta_i$ or $H_i^{\text{free}} = H_i^{\text{Dirac}}$. It is then an obvious idea[1] to consider partial Hamiltonians of the form

$$H_j = H_j^{\text{free}} + V_j(x_1, \ldots, x_N). \tag{2.30}$$

Preferably, the multi-time equations

$$\left(i\frac{\partial}{\partial x_j^0} - H_j\right)\psi = 0 \tag{2.31}$$

should be Lorentz invariant.

---

[1] Even though interaction potentials are usually thought of as arising from the non-relativistic limit of quantum field theories.

We now ask, under which conditions on V_j are the multi-time equations with partial Hamiltonians (2.30) consistent? Here is a first example: if V_j depends only on x_j,

$$H_j = H_j^{\text{free}} + V_j(x_j) \tag{2.32}$$

with $V_j : \mathbb{R}^4 \to \mathbb{R}$, then the consistency condition (2.8) is satisfied, since the particles do not interact. We also need to consider a more general situation without interaction.

Definition 2.1 We call the family $(H_1, \ldots, H_N)$ *non-interacting* if there is a function $\theta : \mathbb{R}^{4N} \to \mathbb{R}$ such that the "gauge transformation"

$$\widetilde{\psi}(x_1, \ldots, x_N) := e^{i\theta(x_1,\ldots,x_N)} \psi(x_1, \ldots, x_N) \tag{2.33}$$

leads to a function $\widetilde{\psi}$ that satisfies the (obviously non-interacting) equations

$$i \frac{\partial \widetilde{\psi}}{\partial x_j^0} = \left(H_j^{\text{free}} + \widetilde{V}_j(x_j) \right) \widetilde{\psi} \tag{2.34}$$

for all $j = 1, \ldots, N$ for some (transformed potential) functions $\widetilde{V}_j : \mathbb{R}^4 \to \mathbb{R}$.

In other words, some $V_j(x_1, \ldots, x_N)$ might look interacting because they depend on $x_1, \ldots, x_N$ (and not just x_j), but they really just lead to a phase factor of an otherwise non-interacting wave function; see also Exercise 2.3. Those potentials we call non-interacting as well. With that, we can formulate our first result.

Theorem 2.3 ([42]) *Let H_j^{free} be the free Dirac operator and $V_j : \mathbb{R}^{4N} \to \mathbb{R}$ smooth. Then the consistency condition (2.8) holds if and only if the family $(H_1, \ldots, H_N)$ is non-interacting.*

Thus, for multi-time theories, interaction by potentials is ruled out, at least for some class of potentials.

Proof Suppose (2.8) is satisfied. Then a direct computation yields that

$$
\begin{aligned}
0 &= \left[H_i, H_j \right] - i \frac{\partial H_j}{\partial t_i} + i \frac{\partial H_i}{\partial t_j} \\
&= \left[H_i^{\text{free}}, V_j \right] + \left[V_i, H_j^{\text{free}} \right] - i \frac{\partial V_j}{\partial t_i} + i \frac{\partial V_i}{\partial t_j} \\
&= -i \sum_{a=1}^{3} \left(\alpha_i^{(a)} \frac{\partial V_j}{\partial x_i^a} - \alpha_j^{(a)} \frac{\partial V_i}{\partial x_j^a} \right) - i \left(\frac{\partial V_j}{\partial t_i} - \frac{\partial V_i}{\partial t_j} \right).
\end{aligned}
\tag{2.35}
$$

Since among all $4^N \times 4^N$ matrices, $\mathbb{1}, \alpha_i^{(a)}, \alpha_j^{(a)}$ for all a (and $i \neq j$) are linearly independent, their prefactors must vanish individually, meaning

$$\frac{\partial V_j}{\partial t_i} = \frac{\partial V_i}{\partial t_j} \quad \text{and} \quad V_j = V_j(\boldsymbol{x}_j, t_1, \ldots, t_N). \tag{2.36}$$

From the first relation we can conclude further that

$$V_j = \tilde{V}_j(\boldsymbol{x}_j, t_j) + \frac{\partial \theta(t_1, \ldots, t_N)}{\partial t_j} \tag{2.37}$$

for some function $\theta(t_1, \ldots, t_N)$. Since the last term can be removed through a gauge transformation, this proves the theorem. $\qquad\square$

Theorem 2.3 has been generalized in several ways in [42]: Interaction potentials are still ruled out if

- we demand the consistency condition only on the interior $\overset{\circ}{\mathscr{S}}_N$ of the set $\mathscr{S}_N$ of spacelike configurations;
- H_j^{free} are arbitrary first-order differential operators with matrix-valued coefficients,

$$H_j^{\text{free}} = -i \sum_{a=1}^{3} A_j^a(x_j) \frac{\partial}{\partial x_j^a} + B_j(x_j), \tag{2.38}$$

  provided $\mathbb{1}$ and the A_j^a are still linearly independent and A_j^a and B_j are smooth functions of x_j;
- V_j is a matrix acting on the jth spin space;
- or H_j^{free} is a second-order differential operator (such as $-\Delta$).

Nickel and Deckert [6] found examples of interacting potentials for which the multi-time equations are consistent. They are matrix-valued, involve a non-trivial spin dependence, and are actually not physically plausible. We will look at such potentials in Exercise 2.4 in Sect. 2.4. Moreover, they violate Poincaré invariance, and the following theorem shows that all Poincaré invariant potentials are ruled out for multi-time evolutions.

Theorem 2.4 ([6]) *Let $V_j \in C^1(\mathscr{S}_N, \mathbb{C}^{K \times K})$. If there is a solution $\psi \in C^2(\mathscr{S}_N, \mathbb{C}^K)$ to the multi-time equations (1.44) with partial Hamiltonians (2.30) for all smooth initial data $\psi((0, \boldsymbol{x}_1), \ldots, (0, \boldsymbol{x}_N))$ with compact support, then V_j is not Poincaré invariant.*

Sketch of proof After evaluating the consistency condition and using linear independence, one finds that V_1 is a combination of $\mathbb{1}_2$ and γ_2^5 (where

$$\gamma^5 := i\gamma^0\gamma^1\gamma^2\gamma^3 \tag{2.39}$$

is a fifth gamma matrix). Translation invariance then leads to explicit expressions for V_1, V_2 that are not Lorentz invariant. $\qquad\square$

2.3 Short Range Interactions

As interaction potentials do not provide consistent multi-time equations, we need alternative ways of implementing interaction. Crater and Van Alstine [5, 60] proposed, instead of potentials V, other interaction operators that are pseudodifferential operators (and not multiplication operators), see also [26]; we will not consider them here. Another natural idea is to use point interactions that occur only when $x_i = x_j$ and are formulated using boundary conditions, see Chap. 3. An obstacle here is that point interactions do not exist for the Dirac equation in 1+3 dimensions [49]: The only self-adjoint Hamiltonian that acts in the region $(\mathbb{R}^3)^N \setminus \{(x_1, \ldots, x_N) : x_i = x_j \text{ for some } i \neq j\}$ like the free Dirac Hamiltonian,

$$i\frac{\partial \varphi}{\partial t} = \sum_{j=1}^{N} H_j^{\text{Dirac}} \varphi, \qquad (2.40)$$

is the free Dirac Hamiltonian itself, so that (2.40) holds on all of $(\mathbb{R}^3)^N$. In other words, there is no point-interaction for the Dirac equation in $1 + 3$ dimensions. However, point interactions exist for the Dirac equation in 1+1 dimensions, and this will be used in Chap. 3. (In general, whether or not there are self-adjoint extensions for Schrödinger or Dirac-type free Hamiltonians depends on the dimension and the order of the differential operator.)

Another idea is to introduce a cut-off length $\delta > 0$ and approximate the point interaction by a short-range interaction of range δ. That is, we allow an interaction potential $W(x_i - x_j)$ such that

$$W(x) = 0 \text{ whenever } |x| \geq \delta. \qquad (2.41)$$

The multi-time wave function is then defined, not on the set $\mathscr{S}$ of spacelike configurations but on the set $\mathscr{S}_\delta$ of "δ-spacelike configurations," i.e., space-time configurations $(x_1, \ldots, x_N) \in \mathscr{M}^N$ such that for every $i \neq j$,

$$\text{either } t_i = t_j \text{ or } |x_i - x_j| > |t_i - t_j| + \delta \qquad (2.42)$$

(see Fig. 2.3).

For any such configuration, we can group together the particles with equal time coordinate and define in this way a partition $S_1 \cup \ldots \cup S_L = \{1, \ldots, N\}$ ($S_\alpha \cap S_\beta =$

Fig. 2.3 A δ-spacelike configuration

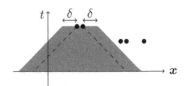

Fig. 2.4 Geometry relevant
to the partial time evolution
from t_{L-1} to t_L

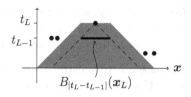

Ø for $\alpha \neq \beta$). We write t_α for the common time value of the family of particles indexed by S_α. Then multi-time equations can be formulated at $(x_1, \ldots, x_N)$ by evolving each group of particles along t_α, i.e.,

$$i\frac{\partial \psi}{\partial t_\alpha} = \left(\sum_{j \in S_\alpha} H_j^{\text{Dirac}} + \frac{1}{2} \sum_{\substack{i,j \in S_\alpha \\ i \neq j}} W(x_i - x_j) \right) \psi \tag{2.43}$$

for all $\alpha = 1, \ldots, L$. Indeed, we find the following result.

Theorem 2.5 ([42]) *The Eq. (2.43) are consistent for smooth W with range δ as in* (2.41).

Note that the set $\mathscr{S}_\delta$ and the Eq. (2.43) are not Lorentz invariant, and in the limit $\delta \to 0$ the equations become non-interacting. Thus, we regard this result less as fundamentally important and more as a remark on the "next-best thing to point-interactions" for the Dirac Hamiltonian in $1 + 3$ dimensions.

Sketch of proof We proceed by induction in L. The start is provided by the one-time evolution. For the induction step, the propagation locality of the Dirac equation is crucial: since the wave function propagates no faster than light, its value at (say) (t_L, x_L) is determined by initial data at time (say) t_{L-1} on the 3-ball $B_{|t_L - t_{L-1}|}(x_L)$ centered at x_L with radius $|t_L - t_{L-1}|$, see Fig. 2.4.

This allows us to define a function ψ; then we use the consistency condition to show that ψ satisfies the Eq. (2.43).   □

2.4   Exercises

Exercise 2.1 (*No-go theorem for potentials in multi-time equations with Laplacians*) Consider the multi-time system

$$\begin{aligned} i\partial_{t_1} \psi &= (-\Delta_1 + V_1(x_1, x_2))\psi, \\ i\partial_{t_2} \psi &= (-\Delta_2 + V_2(x_1, x_2))\psi \end{aligned} \tag{2.44}$$

for a multi-time wave function $\psi : \mathbb{R}^4 \times \mathbb{R}^4 \to \mathbb{C}$. Here, Δ_i denotes the Laplacian with respect to x_i, $i = 1, 2$ and $V_1, V_2 : \mathbb{R}^6 \to \mathbb{R}$ are smooth functions.

(a) State the appropriate consistency condition.
(b) Show that this consistency condition is only satisfied if V_1 does not depend on
     x_2 and V_2 does not depend on x_1.

Exercise 2.2 (*Poincaré invariant interaction potential in multi-time Dirac equations*) Consider the Poincaré invariant multi-time equations

$$\left(i\gamma_k^\mu \partial_{x_k^\mu} - m_k - \frac{e^2}{2\sqrt{|(x_1 - x_2)^2|}} \right) \psi(x_1, x_2) = 0, \quad k = 1, 2, \tag{2.45}$$

where $(x_1 - x_2)^2 = (x_1^0 - x_2^0)^2 - |\mathbf{x}_1 - \mathbf{x}_2|^2$.

(a) Demonstrate that the single-time wave function $\varphi(t, \mathbf{x}_1, \mathbf{x}_2) = \psi(t, \mathbf{x}_1, t, \mathbf{x}_2)$
     satisfies a Schrödinger-like equation with a potential $\propto \frac{e^2}{|\mathbf{x}_1 - \mathbf{x}_2|}$.
(b) Write down the appropriate consistency condition for (2.45).
(c) Show through an explicit calculation that the consistency condition is violated.

Exercise 2.3 (*gauge transformations*) Consider the multi-time equations

$$i\partial_{t_k}\psi = (H_k^0 + V_k)\psi \tag{2.46}$$

for $j = 1, \ldots, N$, where H_k^0 is the free Dirac Hamiltonian

$$H_k^0 = -i \sum_{a=1}^3 \gamma_k^0 \gamma_k^a \partial_{x_k^a} + \gamma_k^0 m_k \tag{2.47}$$

and $V_k \in C^\infty(\mathbb{R}^{4N}, \mathbb{R})$. For some $\theta \in C^\infty(\mathbb{R}^{4N}, \mathbb{R})$ we define the gauge transformation

$$\widetilde{\psi}(x_1, \ldots, x_N) = e^{i\theta(x_1, \ldots, x_N)} \psi(x_1, \ldots, x_N). \tag{2.48}$$

Compute which equations $\widetilde{\psi}$ satisfies.

Note: The result will show that some potentials in the Dirac equation do not really lead to interaction but are only due to gauge effects.

Exercise 2.4 (*consistent interaction potentials*) Consider the multi-time equations

$$\begin{aligned} i\partial_{t_1}\psi &= (H_1^0 + V_1)\psi, \\ i\partial_{t_2}\psi &= (H_2^0 + V_2)\psi, \end{aligned} \tag{2.49}$$

where H_k^0 is the free Dirac Hamiltonian $H_k^0 = -i \sum_{a=1}^3 \gamma_k^0 \gamma_k^a \partial_{x_k^a} + \gamma_k^0 m_k$. We now consider the matrix valued potentials (see [6])

$$\begin{aligned} V_1(x_1, x_2) &= \gamma_1^\mu C_\mu \exp\left(2i\gamma_1^5 c_\lambda (x_2^\lambda - x_1^\lambda)\right) - m_1\gamma_1^0, \\ V_2 &= \gamma_1^5 \gamma_2^0 \gamma_2^\nu c_\nu, \end{aligned} \tag{2.50}$$

for arbitrary non-zero c, $C \in \mathbb{C}^4$ and $\gamma^5 = i\gamma^0\gamma^1\gamma^2\gamma^3$. Then $\gamma^5\gamma^\mu = -\gamma^\mu\gamma^5$ for $\mu = 0, 1, 2, 3$. Show that the multi-time equations (2.49) are consistent.

Note: It can be shown that these multi-time equations are indeed interacting, i.e., there is no gauge transformation which makes them non-interacting. However, they are not Poincaré invariant.

Chapter 3
Relativistic Point Interactions in 1+1 Dimensions

The model we will present in this chapter [25, 27, 29] was the first rigorous example showing that an interacting, Lorentz invariant dynamics is possible for multi-time wave functions. It involves point interaction (also known as delta interaction or zero-range interaction) that has an effect only where two particles meet. The interaction is mathematically implemented via a boundary condition, and the dynamics is otherwise free, which helps obey the consistency condition.

3.1 Setting

We will describe the model for two Dirac particles (the physically most relevant case) in 1+1 space-time dimensions (then point interaction is feasible mathematically) in the massless case (then the Dirac equation is easily solvable explicitly).

3.2 Definition of the Model

The multi-time wave function is of the form

$$\psi : \overset{\circ}{\mathscr{S}_1} \subset \mathbb{R}^2 \times \mathbb{R}^2 \to \mathbb{C}^2 \otimes \mathbb{C}^2 \cong \mathbb{C}^4 ,$$

$$\underbrace{(t_1, z_1,}_{x_1} \underbrace{t_2, z_2)}_{x_2} \mapsto \begin{pmatrix} \psi_{--} \\ \psi_{-+} \\ \psi_{+-} \\ \psi_{++} \end{pmatrix} (t_1, z_1, t_2, z_2) . \qquad (3.1)$$

In 1+1 dimensions, the interior $\overset{\circ}{\mathscr{S}}$ of the set $\mathscr{S}$ of spacelike configurations is the disjoint union of two connected components $\overset{\circ}{\mathscr{S}_1}$ and $\overset{\circ}{\mathscr{S}_2}$ given by (see Fig. 3.1)

© The Author(s), under exclusive license to Springer Nature Switzerland AG 2020
M. Lienert et al., *Multi-time Wave Functions*,
SpringerBriefs in Physics, https://doi.org/10.1007/978-3-030-60691-6_3

Fig. 3.1 The interiors of the
sets $\mathscr{S}_i$ as in (3.2) in relative
coordinates

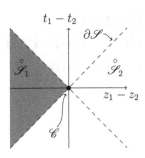

$$\overset{\circ}{\mathscr{S}}_{1/2} = \left\{ (t_1, z_1, t_2, z_2) \in \mathbb{R}^2 \times \mathbb{R}^2 : |t_1 - t_2| < |z_1 - z_2| \text{ and } z_1 \lessgtr z_2 \right\}. \quad (3.2)$$

As a consequence, it suffices to set up the model on the reduced domain $\overset{\circ}{\mathscr{S}}_1$ (where $z_1 < z_2$). The relevant boundary is not all of $\partial \overset{\circ}{\mathscr{S}}_1$ but just the set $\mathscr{C}$ of coincidence points,

$$\partial \overset{\circ}{\mathscr{S}}_1 \supset \mathscr{C} = \left\{ (t_1, z_1, t_2, z_2) \in \mathbb{R}^2 \times \mathbb{R}^2 : t_1 = t_2 \text{ and } z_1 = z_2 \right\}. \quad (3.3)$$

The multi-time equations are the free Dirac equations on $\overset{\circ}{\mathscr{S}}_1$,

$$i\gamma_k^{\mu} \partial_{k,\mu} \psi(x_1, x_2) = 0 \quad \text{for } k = 1, 2 \quad (3.4)$$

with $\partial_{k,\mu} = \frac{\partial}{\partial x_k^{\mu}}$ and $\mu = 0, 1$ summed over. In the representation we choose,

$$\gamma^0 = \sigma^1 = \begin{pmatrix} 0 & 1 \\ 1 & 0 \end{pmatrix}, \quad \gamma^1 = \sigma^1 \sigma^3 \text{ with } \sigma^3 = \begin{pmatrix} 1 & 0 \\ 0 & -1 \end{pmatrix}. \quad (3.5)$$

The initial conditions are of the form

$$\psi(0, z_1, 0, z_2) = \psi_0(z_1, z_2) \quad \text{for } z_1 < z_2 \quad (3.6)$$

with a given C^1 function ψ_0.

The boundary conditions are a linear relation between the components of ψ on $\mathscr{C}$. They can be written as

$$M\psi(t, z, t, z) = 0 \quad (3.7)$$

with $\psi(t, z, t, z)$ the limit of ψ towards the boundary, and the detailed form of the 1×4-matrix M to be chosen later (Sect. 3.4) in such a way that probability is locally conserved.

3.3 General Solution

We multiply (3.4) by γ_k^0 from the left (note that $(\gamma^0)^2 = \mathbf{1}$) to obtain

$$(\partial_{t_k} + \sigma_k^3 \partial_{z_k})\psi = 0 . \tag{3.8}$$

As $\sigma^3 = \begin{pmatrix} 1 & 0 \\ 0 & -1 \end{pmatrix}$, this equation is diagonal. This is best seen using the notation $\psi = \psi_{s_1 s_2}$ with $s_i = \mp 1$, in which we can rewrite (3.4) as

$$(\partial_{t_k} - s_k \partial_{z_k})\psi_{s_1 s_2} = 0 \quad \text{for } k = 1, 2 . \tag{3.9}$$

Examples:

$$(\partial_{t_1} + \partial_{z_1})\psi_{--} = 0 \tag{3.10a}$$
$$(\partial_{t_2} + \partial_{z_2})\psi_{--} = 0 \tag{3.10b}$$
$$(\partial_{t_1} - \partial_{z_1})\psi_{+-} = 0 \tag{3.10c}$$
$$(\partial_{t_2} + \partial_{z_2})\psi_{+-} = 0 \quad \text{etc.} \tag{3.10d}$$

Solutions of (3.10a), (3.10b) have the form

$$\psi_{--} = f_{--}(z_1 - t_1, z_2 - t_2) \tag{3.11}$$

for some C^1 function f_{--}. In general,

$$\psi_{s_1 s_2} = f_{s_1 s_2}(z_1 + s_1 t_1, z_2 + s_2 t_2) . \tag{3.12}$$

The upshot is that if we know $f_{s_1 s_2}$, we know $\psi_{s_1 s_2}$; $f_{s_1 s_2}$ should be determined via the initial conditions or the boundary conditions. Note also that $\psi_{s_1 s_2}$ is constant along the following 2d surfaces in $\mathbb{R}^2 \times \mathbb{R}^2$ ("multi-time characteristics")

$$z_1 + s_1 t_1 = \text{const.}, \quad z_2 + s_2 t_2 = \text{const.} \tag{3.13}$$

So, if we want to know ψ in $p = (t_1, z_1, t_2, z_2)$, then we have to see where the respective multi-time characteristic intersects a surface where ψ is already known.

The first candidate is the initial data surface $t_1 = t_2 = 0$. We have that

$$\psi_{s_1 s_2}(0, z_1, 0, z_2) = f_{s_1 s_2}(z_1 + 0, z_2 + 0) = \psi_{0, s_1 s_2}(z_1, z_2) \quad \text{for } z_1 < z_2. \tag{3.14}$$

This determines $f_{s_1 s_2}(x, y)$ for $x < y$, i.e., half of the values of f.

Do we need more values of $f_{s_1 s_2}$? We consider the component of ψ separately for $(t_1, z_1, t_2, z_2) \in \overset{\circ}{\mathscr{S}_1}$, i.e., $z_1 < z_2$ and $|t_1 - t_2| < |z_1 - z_2|$:

(a) $\psi_{--} = f_{--}(z_1 - t_1, z_2 - t_2)$: Can $z_1 - t_1 < z_2 - t_2$ occur? It occurs whenever $t_2 - t_1 > z_2 - z_1 = |z_1 - z_2|$, which implies $|t_1 - t_2| > |z_1 - z_2|$, so this case cannot occur. Similarly for ψ_{++}.

(b) $\psi_{-+} = f_{-+}(z_1 - t_1, z_2 + t_2)$. Can $z_1 - t_1 < z_2 + t_2$ occur? Consider, e.g., $t_1 = t_2 = t$. Then (t, z_1, t, z_2) is always spacelike, and $z_1 - t > z_2 + t$ if and only if $-t > (z_2 - z_1)/2 > 0$. This can occur for negative times t! Similarly, ψ_{+-} is not determined for $(t_1 + t_2) > z_2 - z_1 > 0$.

This suggests that ψ_{--} and ψ_{++} should not be subject to a boundary condition, whereas ψ_{-+} and ψ_{+-} should (alternatingly for positive/negative times).

How does a boundary condition work (say, for $t < 0$)? Suppose we impose the boundary values

$$\psi_{-+}(t, z, t, z) = h_1(t, z) \tag{3.15}$$

for a given function h_1. Then

$$f_{-+}(z - t, z + t) = h_1(t, z), \tag{3.16}$$

which conversely determines f_{-+} according to

$$f_{-+}(a, b) = h_1 \left(\tfrac{b-a}{2}, \tfrac{b+a}{2} \right). \tag{3.17}$$

But how to choose these boundary conditions? We need physical considerations!

3.4   Boundary Conditions from Local Probability Conservation

Probability conservation for a domain $\Omega \subseteq \mathbb{R}^2 \times \mathbb{R}^2$ means that

$$P(\Sigma) = P(\Sigma') \tag{3.18}$$

for all Cauchy surfaces Σ, Σ', where

$$P(\Sigma) = \int\limits_{(\Sigma \times \Sigma) \cap \Omega} d\sigma_{1\mu} \, d\sigma_{2\nu} \, \underbrace{\overline{\psi} \gamma_1^\mu \gamma_2^\nu \psi}_{j^{\mu\nu}} \tag{3.19}$$

and $d\sigma_\mu$ is short for $d^3\sigma(x) \, n_\mu(x)$. For multi-time equations on $\Omega = \mathbb{R}^2 \times \mathbb{R}^2$, we know already that this holds, as it follows from

$$\partial_{1\mu} j^{\mu\nu} = 0 = \partial_{2\nu} j^{\mu\nu}, \tag{3.20}$$

Fig. 3.2 Visualization of the set S_R for equal times in the case that Σ and Σ' are horizontal

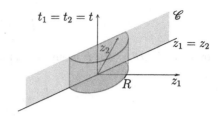

true by (1.56), via the divergence theorem, see Exercise 1.4. Here, for $\Omega = \overset{\circ}{\mathscr{S}}_1$, we need to consider that probability can get lost through the boundary.

As we will show, one needs to avoid a loss of probability through $\mathscr{C}$ (the only relevant boundary, as $(\Sigma \times \Sigma) \cap \mathscr{S}_1$ only has boundary in $\mathscr{C}$). This leads to

$$(j^{01} - j^{10})(t, z, t, z) = 0. \tag{3.21}$$

Here is the more detailed reasoning using Stokes' theorem. Let ψ by compactly supported in space, i.e., $\psi(t_1, z_1, t_2, z_2) = 0$ if $|z_1| > R$ or $|z_2| > R$ for some $R > 0$. Define

$$\Sigma_R := \{(t, z) \in \Sigma : |z| < R\} \tag{3.22}$$

and analogously Σ'_R. One can construct a closed 2-dimensional surface $S_R \subset \mathbb{R}^2 \times \mathbb{R}^2$ of the form

$$S_R = \left[(\Sigma_R \times \Sigma_R) \cap \overset{\circ}{\mathscr{S}}_1\right] \cup \left[(\Sigma'_R \times \Sigma'_R) \cap \overset{\circ}{\mathscr{S}}_1\right] \cup M_R \cup M_\mathscr{C} \tag{3.23}$$

with M_R a "mantle" surface contained in $\{|z_1| = R = |z_2|\}$ (so that ψ vanishes on M_R) and $M_\mathscr{C}$ another mantle surface contained in $\mathscr{C}$; see Fig. 3.2.

Now we define the differential form (see [38, Sect. 2.5] for an introduction to differential forms)

$$\omega_j = d\sigma_{1\mu} d\sigma_{2\nu} j^{\mu\nu}, \tag{3.24}$$

the *current density 2-form*, that is,

$$\omega_j = \sum_{\mu,\nu=0}^{1} (-1)^{\mu+\nu} j^{\mu\nu} dx_1^{1-\mu} \wedge dx_2^{1-\nu}$$

$$= j^{00} dx_1^1 \wedge dx_1^1 - j^{01} dx_1^1 \wedge dx_2^0 - j^{10} dx_1^0 \wedge dx_2^1 + j^{11} dx_1^0 \wedge dx_2^0. \tag{3.25}$$

Exercise 3.1 Show that the relation $\partial_{j\mu_j} j^{\mu_1\mu_2} = 0$ is equivalent to $d\omega_j = 0$, where d means the exterior derivative of a differential form.

Stokes' theorem asserts that for any differential r-form ω and any $r + 1$-dimensional surface F,

$$\int_F d\omega = \int_{\partial F} \omega \, . \tag{3.26}$$

In our case, $\partial F = S_R$, F is any 3-volume enclosed by S_R, and $\omega = \omega_j$, so

$$\int_{S_R} \omega_j = \int_F d\omega_j = 0 \, . \tag{3.27}$$

Thus,

$$\int_{(\Sigma_R \times \Sigma_R) \cap \mathring{\mathscr{S}}_1} \omega_j = \underbrace{\int_{(\Sigma_R \times \Sigma_R) \cap \mathring{\mathscr{S}}_1} \omega_j}_{} + \underbrace{\int_{M_R} \omega_j}_{=0 \text{ as } \psi|_{M_R}=0} + \int_{M_{\mathscr{C}}} \omega_j \, . \tag{3.28}$$

Therefore, probability conservation is equivalent to

$$\int_{M_{\mathscr{C}}} \omega_j = 0 \, . \tag{3.29}$$

Our boundary condition needs to ensure that. We demand that

$$\omega_j \Big|_{\mathscr{C}} = 0 \, . \tag{3.30}$$

We evaluate this condition in relative coordinates $z = z_1 - z_2$, $Z = z_1 + z_2$, $\tau = t_1 - t_2$, $T = t_1 + t_2$:

$$\begin{aligned}
\omega_j = {} & \tfrac{1}{2} j^{00} \, dz \wedge dZ - \tfrac{1}{4}(j^{10} + j^{01}) d\tau \wedge dZ + \tfrac{1}{4}(j^{10} - j^{01}) d\tau \wedge dz \\
& - \tfrac{1}{4}(j^{10} - j^{01}) dT \wedge dZ - \tfrac{1}{4}(j^{10} + j^{01}) dz \wedge dT + \tfrac{1}{2} j^{11} \, d\tau \wedge dT \, .
\end{aligned} \tag{3.31}$$

On $\mathscr{C}$, $z = 0 = \tau$, so

$$\omega_j \Big|_{\mathscr{C}} = -\tfrac{1}{4}(j^{10} - j^{01}) dT \wedge dZ \, , \tag{3.32}$$

and this vanishes if and only if

$$(j^{10} - j^{01})(t, z, t, z) = 0 \quad \forall t, z \in \mathbb{R} \, . \tag{3.33}$$

This is our condition for local probability conservation. How do we convert this into a boundary condition for ψ? If we write out $j^{\mu\nu} = \bar{\psi}\gamma_1^\mu \gamma_2^\nu \psi$ in components of ψ,

$$j^{\mu\nu} = |\psi_{--}|^2 + (-1)^\nu |\psi_{-+}|^2 + (-1)^\mu |\psi_{+-}|^2 + (-1)^{\mu+\nu} |\psi_{++}|^2 \, , \tag{3.34}$$

then

$$j^{10} - j^{01} = 2\big(|\psi_{+-}|^2 - |\psi_{-+}|^2\big) \, , \tag{3.35}$$

and (3.33) is equivalent to

$$\psi_{-+} = e^{i\theta}\psi_{+-} \quad \text{on } \mathscr{C} \tag{3.36}$$

for some $\theta \in [0, 2\pi)$. In principle, θ could be a function of t and z, but then the dynamics would not be translation invariant. So (3.36) is our boundary condition!

3.5 Main Result

With our previous considerations, one can show the following theorem; a more detailed proof can be found in the paper [25].

Theorem 3.1 *Let $\theta \in [0, 2\pi)$. Our model, defined by*

$$i\gamma_k^\mu \partial_{k\mu}\psi = 0 \quad (k = 1, 2) \tag{3.37a}$$

$$\psi(0, z_1, 0, z_2) = \psi_0(z_1, z_2) \quad (z_1 < z_2) \tag{3.37b}$$

$$\psi_{-+} = e^{i\theta}\psi_{+-} \quad \text{on } \mathscr{C} \tag{3.37c}$$

has a unique solution for every $\psi_0 \in C^1(\mathbb{R}^2)$.

The solution can be specified explicitly: it is given by

$$\begin{pmatrix} \psi_{--} \\ \psi_{-+} \\ \psi_{+-} \\ \psi_{++} \end{pmatrix}(t_1, z_1, t_2, z_2) = \begin{pmatrix} \psi_{0--}(z_1 - t_1, z_2 - t_2) \\ \begin{cases} \psi_{0-+}(z_1 - t_1, z_2 + t_2) & \text{if } z_1 - t_1 \le z_2 + t_2 \\ \psi_{0+-}(z_2 + t_2, z_1 - t_1) & \text{else} \end{cases} \\ \begin{cases} \psi_{0+-}(z_1 + t_1, z_2 - t_2) & \text{if } z_1 + t_1 \le z_2 - t_2 \\ \psi_{0-+}(z_2 - t_2, z_1 + t_1) & \text{else} \end{cases} \\ \psi_{0++}(z_1 + t_1, z_2 + t_2) \end{pmatrix}. \tag{3.38}$$

The solution is continuous if ψ_0 satisfies the boundary condition and continuously differentiable if in addition

$$(\partial_1\psi_{0-+})(z, z) = e^{i\theta}(\partial_2\psi_{0+-})(z, z), \tag{3.39a}$$

$$(\partial_2\psi_{0-+})(z, z) = e^{i\theta}(\partial_1\psi_{0+-})(z, z). \tag{3.39b}$$

Moreover, if ψ_0 is compactly supported, then ψ is compactly supported in space for all t_1, t_2, and probability is conserved in the sense that $P(\Sigma) = P(\Sigma')$ for all Cauchy surfaces $\Sigma, \Sigma' \subset \mathbb{R}^2$ and $P(\Sigma)$ defined in (3.19).

3.6 Lorentz Invariance

Except for the boundary condition, Lorentz invariance is already manifest. To check the Lorentz invariance of the boundary condition, note that for a proper Lorentz transformation $\Lambda \in \mathscr{L}^+$,

$$\psi'(t'_1, z'_1, t'_2, z'_2) = S[\Lambda] \otimes S[\Lambda]\,\psi(t_1, z_1, t_2, z_2). \tag{3.40}$$

The action $S[\Lambda]$ of Λ on spinors can be computed explicitly by evaluating the following matrix exponential if Λ is a boost with rapidity (Minkowski angle) $\omega \in \mathbb{R}$:

$$
\begin{aligned}
S[\Lambda] &= \exp\left(\frac{\omega}{2}\gamma^0\gamma^1\right) \\
&= \cosh(\omega/2)\,\mathbb{1}_2 + \sinh(\omega/2)\,\sigma^3 \\
&= \begin{pmatrix} \cosh(\omega/2) + \sinh(\omega/2) & 0 \\ 0 & \cosh(\omega/2) - \sinh(\omega/2) \end{pmatrix}.
\end{aligned} \tag{3.41}
$$

From this it follows that

$$\psi'_{-+} = \big(\cosh(\omega/2) + \sinh(\omega/2)\big)\big(\cosh(\omega/2) - \sinh(\omega/2)\big)\psi_{-+} = \psi_{-+} \tag{3.42a}$$

$$\psi'_{+-} = \big(\cosh(\omega/2) - \sinh(\omega/2)\big)\big(\cosh(\omega/2) + \sinh(\omega/2)\big)\psi_{+-} = \psi_{+-}. \tag{3.42b}$$

Thus, as (t', z', t', z') is again in $\mathscr{C}$, the boundary condition is indeed Lorentz invariant!

3.7 Upshot

Interaction. The goal was to construct an *interacting* model. How do we check that? The meaning of *interaction* is that there are some initial product wave functions which get entangled with time. In contrast, "free" would mean that every initial product wave function remains a product wave function also for later times. Our model here is indeed interacting, as one can see from the solution formula.

Conclusion. The model presented indeed provides the first rigorous example of a manifestly Lorentz invariant, interacting multi-time model compatible with a probabilistic interpretation.

Outlook.

- The case of N particles has also been treated [29].
- We assumed $m = 0$. The case with $m > 0$ is certainly also possible (for the same boundary conditions) but a bit more indirect due to the more complicated solution of the Dirac equation.
- A model for a simple quantum field theory in a similar spirit will be discussed in Chap. 5.

- A version in higher dimension is probably not possible in the same way because, as discussed in Sect. 2.3, it is known [49] that point interaction is impossible for a Dirac particle in 2 or 3 dimensions.
- A model with contact interactions between an electron and a photon in 1 dimension is treated in [23].

3.8 Exercises

Exercise 3.2 (*δ-potential in a 2-particle Dirac equation*) Consider the single-time Dirac equation for two massless particles in 1+1 dimensions interacting through a δ-potential:

$$i\partial_t \varphi(t, z_1, z_2) = - i \left(\sigma^3 \otimes 1_2 \, \partial_{z_1} + 1_2 \otimes \sigma^3 \, \partial_{z_2}\right) \varphi(t, z_1, z_2) \\ + M \, \delta(z_1 - z_2)\varphi(t, z_1, z_2). \tag{3.43}$$

Here, $\sigma^3 \otimes 1_2 = \mathrm{diag}(1, 1, -1, -1)$, $\sigma^3 \otimes 1_2 = \mathrm{diag}(1, -1, 1, -1)$ and $M = \mathrm{diag}(0, 1, -1, 0)$ in the basis corresponding to $\varphi = (\varphi_{--}, \varphi_{-+}, \varphi_{+-}, \varphi_{++})$. Assume that the spin components φ_{--} and φ_{++} are continuous across $z_1 = z_2$ but that φ_{-+} and φ_{+-} may have a discontinuity across that line.

(a) By integrating over (3.43) in the relative variable $z = z_1 - z_2$ over an ε-neighborhood of $z = 0$, extract a boundary condition at $z_1 = z_2$ in the limit $\varepsilon \to 0$.
(b) Assuming antisymmetry of φ, compare this boundary condition to the multi-time model in this chapter.

Exercise 3.3 (*general boundary conditions for Dirac particles in 1+1 dimensions*) On $\mathscr{S}_1 = \{(t_1, z_1, t_2, z_2) \in \mathbb{R}^4 : |t_1 - t_2| < |z_1 - z_2| \text{ and } z_1 < z_2\}$ we have found the general solution

$$\begin{pmatrix} \psi_{--} \\ \psi_{-+} \\ \psi_{+-} \\ \psi_{++} \end{pmatrix} (t_1, z_1, t_2, z_2) = \begin{pmatrix} f_{--}(z_1 - t_1, z_2 - t_2) \\ f_{-+}(z_1 - t_1, z_2 + t_2) \\ f_{+-}(z_1 + t_1, z_2 - t_2) \\ f_{++}(z_1 + t_1, z_2 + t_2) \end{pmatrix}, \tag{3.44}$$

where $f_{s_1 s_2}$ are C^1-functions whose values $f_{s_1 s_2}(a, b) = g_{s_1 s_2}(a, b)$ are fixed through initial data $g_{s_1 s_2} \in C^1(\mathbb{R}^2, \mathbb{C})$ for $a \leq b$.
Let $h \in C^1(\mathbb{R}^2, \mathbb{C})$ be a given function. Show that together with the initial data, the boundary condition

$$\psi_{+-}(t, z, t, z) = h(t, z), \quad t \geq 0, z \in \mathbb{R} \tag{3.45}$$

determines ψ completely on $\mathscr{S}_1$ and for $t_1, t_2 \geq 0$.

Chapter 4
Multi-time Quantum Field Theory

When we aim for interacting multi-time equations, we saw in the previous two chapters that potentials are mostly ruled out, and point interactions are not available in $3 + 1$ dimensions (for Dirac particles). However, another approach is that interaction is mediated by the creation and annihilation of other particles. This is the physical picture behind quantum field theory (QFT). In this chapter, we explore how multi-time equations with particle creation can be set up. If the number of particles is not fixed, then of course also the number of time variables is not fixed but differs from sector to sector, so we need to rethink the question of how to check consistency. We show on the non-rigorous level for simple QFT models that such equations can indeed be consistent and report about rigorous results. For this reason, particle creation is a natural option for implementing interaction in a multi-time theory.

4.1 Fock Space

Let us first give a brief overview of the Fock space formalism (introductions can be found in many textbooks, e.g., [46]). So far, $\varphi(t, \boldsymbol{x}_1, \dots, \boldsymbol{x}_N) \in (\mathbb{C}^4)^{\otimes N}$ ($=$ the spin space for Dirac particles), with $(\boldsymbol{x}_1, \dots, \boldsymbol{x}_N) \in (\mathbb{R}^3)^N$, the configuration space. Sometimes we will use the notation $\varphi_{s_1 \dots s_N}(t, \boldsymbol{x}_1, \dots, \boldsymbol{x}_N)$, explicitly highlighting the spin indices $s_1, \dots, s_N \in \mathbb{C}^4$. Thus, for each $t \in \mathbb{R}$, $\varphi(t) \in S_\pm L^2(\mathbb{R}^3, \mathbb{C}^4)^{\otimes N}$, the Hilbert space for N particles, with $S_\pm$ the symmetrization/anti-symmetrization operator for bosons/fermions. The scalar product is given by

$$\langle \varphi | \widetilde{\varphi} \rangle_N = \int \mathrm{d}^3 \boldsymbol{x}_1 \cdots \int \mathrm{d}^3 \boldsymbol{x}_N \, \varphi^*(\boldsymbol{x}_1, \dots, \boldsymbol{x}_N) \, \widetilde{\varphi}(\boldsymbol{x}_1, \dots, \boldsymbol{x}_N) . \tag{4.1}$$

Now we introduce a state with variable particle number,

© The Author(s), under exclusive license to Springer Nature Switzerland AG 2020
M. Lienert et al., *Multi-time Wave Functions*,
SpringerBriefs in Physics, https://doi.org/10.1007/978-3-030-60691-6_4

$$\varphi = \begin{pmatrix} \varphi(\emptyset) \\ \varphi(\boldsymbol{x}_1) \\ \varphi(\boldsymbol{x}_1, \boldsymbol{x}_2) \\ \vdots \\ \varphi(\boldsymbol{x}_1, \ldots, \boldsymbol{x}_N) \\ \vdots \end{pmatrix} \qquad (4.2)$$

and call $\varphi(\boldsymbol{x}_1, \ldots, \boldsymbol{x}_N)$ the N-particle sector of φ, also denoted $\varphi^{(N)}$. The corresponding configuration space is

$$\bigcup_{N=0}^{\infty} (\mathbb{R}^3)^N =: \Gamma(\mathbb{R}^3), \qquad (4.3)$$

and

$$\varphi \in \bigoplus_{N=0}^{\infty} S_{\pm} L^2(\mathbb{R}^3, \mathbb{C}^4)^{\otimes N} =: \mathscr{F}_{\pm}, \qquad (4.4)$$

the Fock space. $\mathscr{F}_{\pm}$ is still a Hilbert space equipped with the scalar product

$$\langle \varphi | \tilde{\varphi} \rangle = \sum_{N=0}^{\infty} \langle \varphi^{(N)} | \tilde{\varphi}^{(N)} \rangle_N . \qquad (4.5)$$

Two operators connect between different Fock space sectors: the annihilation operator $a_s(\boldsymbol{x})$,

$$\left(a_s(\boldsymbol{x}) \varphi \right)^{(N)}_{s_1 \ldots s_N} (\boldsymbol{x}_1, \ldots, \boldsymbol{x}_N) = \sqrt{N+1} \, \varepsilon^N \, \varphi^{(N+1)}_{s_1 \ldots s_N, s}(\boldsymbol{x}_1, \ldots, \boldsymbol{x}_N, \boldsymbol{x}), \qquad (4.6)$$

where the left-hand side lies in the N-particle sector, $\sqrt{N+1}$ is a convenient combinatorial factor, ε has the value $+1$ for bosons and -1 for fermions (to preserve anti-symmetry) and $\varphi^{(N+1)}$ lies in the $N+1$-particle sector; and the creation operator $a_s^{\dagger}(\boldsymbol{x})$, the adjoint of $a_s(\boldsymbol{x})$,

$$\left(a_s^{\dagger}(\boldsymbol{x}) \varphi \right)^{(N)}_{s_1 \ldots s_N} (\boldsymbol{x}_1, \ldots, \boldsymbol{x}_N)$$
$$= \frac{1}{\sqrt{N}} \sum_{j=1}^{N} \varepsilon^{j+1} \delta_{ss_j} \delta^3(\boldsymbol{x}_j - \boldsymbol{x}) \varphi^{(N-1)}_{s_1 \ldots \hat{s}_j \ldots s_N}(\boldsymbol{x}_1, \ldots, \widehat{\boldsymbol{x}}_j, \ldots, \boldsymbol{x}_N), \qquad (4.7)$$

where the $\sum_j \varepsilon^{j+1}$ ensures (anti-)symmetrization, $\delta^3(\boldsymbol{x}) = \delta(x^1)\delta(x^2)\delta(x^3)$ means the Dirac delta function in 3 dimensions, the factor $\delta_{ss_j} \delta^3(\boldsymbol{x}_j - \boldsymbol{x})$ plays the role of the "wave function of the created particle," and the hat means omission. Note that another notation for omitting indices is $(s_1 \ldots s_N) \setminus s_j$, and correspondingly for the arguments $(\boldsymbol{x}_1, \ldots, \boldsymbol{x}_N) \setminus \boldsymbol{x}_j$; when only one spin index changes we sometimes write $\varphi_{s_1 \ldots s_N, s} = \varphi_s$ or $\varphi_{s_1 \ldots \hat{s}_j \ldots s_N} = \varphi_{\hat{s}_j}$ for brevity.

The creation and annihilation operators satisfy the canonical (anti-)commutation relations (CCR/CAR)

$$\left[a_s^\dagger(\boldsymbol{x}), a_{s'}^\dagger(\boldsymbol{y}) \right]_\varepsilon = 0 = \left[a_s(\boldsymbol{x}), a_{s'}(\boldsymbol{y}) \right]_\varepsilon, \tag{4.8a}$$

$$\left[a_s(\boldsymbol{x}), a_{s'}^\dagger(\boldsymbol{y}) \right]_\varepsilon = \delta_{ss'} \, \delta^3(\boldsymbol{x} - \boldsymbol{y}), \tag{4.8b}$$

where $[\cdot, \cdot]_{\varepsilon=1}$ means the commutator and $[\cdot, \cdot]_{\varepsilon=-1}$ the anti-commutator,

$$[A, B]_{+1} = AB - BA, \quad [A, B]_{-1} = AB + BA. \tag{4.9}$$

The *number operator* $\mathcal{N}$, defined by

$$(\mathcal{N}\varphi)^{(N)} = N\varphi^{(N)}, \tag{4.10}$$

can be expressed as

$$\mathcal{N} = \int_{\mathbb{R}^3} \mathrm{d}^3 x \sum_{s=1}^{4} a_s^\dagger(\boldsymbol{x}) \, a_s(\boldsymbol{x}). \tag{4.11}$$

The free Hamiltonian on $\mathscr{F}_\pm$ can be constructed from the 1-particle Hamiltonian $H_{\boldsymbol{x}}$ (defined on $L^2(\mathbb{R}^3, \mathbb{C}^K)$ as a differential plus a multiplication operator) by the following scheme (often called "second quantization"):

$$\int_{\mathbb{R}^3} \mathrm{d}^3 x \sum_{s=1}^{4} a_s^\dagger(\boldsymbol{x}) \, H_{\boldsymbol{x}} \, a_s(\boldsymbol{x}) \, \varphi^{(N)} = \sum_{j=1}^{N} H_{\boldsymbol{x}_j} \varphi^{(N)}, \tag{4.12}$$

as we will verify in Exercise 4.2 in Sect. 4.7. The combination

$$\Phi_s(\boldsymbol{x}) = a_s(\boldsymbol{x}) + a_s^\dagger(\boldsymbol{x}) \tag{4.13}$$

is usually called the *field operator* (at least in the bosonic case). $\Phi_s(\boldsymbol{x})$ and $a_s^\dagger(\boldsymbol{x})$ (and $a_s(\boldsymbol{x})$) are often called *operator-valued distributions* (since they involve Dirac delta functions, which mathematicians call "delta distributions" because, strictly speaking, they are not functions of $\boldsymbol{x}$).

4.2 Emission–Absorption Model

Let us consider a simple model, based on QFT models of Landau and Peierls [24], Schweber [46], and Nelson [40], where a fixed number M of fermions called "x-particles" can emit and absorb bosons called "y-particles." Apart from emission and absorption, the evolution is free. For simplicity, we take both x and y as Dirac particles (i.e., the y are bosonic Dirac particles, contrary to the spin-statistics theorem). The corresponding Hilbert space is

$$\mathscr{F}_{x,y} = \mathscr{F}_x \otimes \mathscr{F}_y, \tag{4.14}$$

where $\mathscr{F}_x = \mathscr{F}_-$ and $\mathscr{F}_y = \mathscr{F}_+$. The Schrödinger equation

$$i\frac{\mathrm{d}\varphi}{\mathrm{d}t} = H\varphi(t) \tag{4.15}$$

involves the Hamiltonian

$$H = \underbrace{\int \mathrm{d}^3x \sum_{r,r'=1}^{4} a_r^\dagger(x) H_x^{\mathrm{Dirac},r,r'} a_{r'}(x)}_{H_x} + \underbrace{\int \mathrm{d}^3y \sum_{s,s'=1}^{4} b_s^\dagger(y) H_y^{\mathrm{Dirac},s,s'} b_{s'}(y)}_{H_y}$$

$$+ \underbrace{\int \mathrm{d}^3x \sum_{r,s=1}^{4} a_r^\dagger(x) \big(g_s^* b_s(x) + g_s b_s^\dagger(x)\big) a_r(x)}_{H_{\mathrm{int}}}, \tag{4.16}$$

where a is the x annihilation operator, b the y annihilation operator, and $g \in \mathbb{C}^4$ a given constant spinor. Explicitly, in each sector, abbreviating $(x_1, \ldots, x_M) = x^{3M}$ and $(y_1, \ldots, y_N) = y^{3N}$,

$$i\frac{\partial\varphi}{\partial t}(x^{3M}, y^{3N}) = (H\varphi)(x^{3M}, y^{3N})$$

$$= \sum_{j=1}^{M} H_{x_j}^{\mathrm{Dirac}}\varphi(x^{3M}, y^{3N}) + \sum_{k=1}^{N} H_{y_k}^{\mathrm{Dirac}}\varphi(x^{3M}, y^{3N})$$

$$+ \sqrt{N+1} \sum_{j=1}^{M} \sum_{s_{N+1}=1}^{4} g_{s_{N+1}}^* \varphi_{s_{N+1}}\big(x^{3M}, (y^{3N}, x_j)\big)$$

$$+ \frac{1}{\sqrt{N}} \sum_{j=1}^{M} \sum_{k=1}^{N} g_{s_k} \delta^3(y_k - x_j)\varphi_{\widehat{s_k}}\big(x^{3M}, (y^{3N} \setminus y_k)\big). \tag{4.17}$$

This equation expresses the change in the N-particle sector related to the $(N + 1)$- and $(N - 1)$-particle sectors. It is clearly interacting; in fact, for the very similar *Nelson model* [40], non-relativistic or adiabatic limits lead to two-body Coulomb $(1/r)$ or Yukawa (e^{-r}/r) interaction potentials (see [50] and references therein). The model can likewise be set up in curved space-time. Note that this model is mathematically ill-defined because it is ultraviolet divergent. In order to deal with this problem, one may introduce an ultraviolet cut-off (i.e., replace $\delta^3(x)$ by a suitable function $\Lambda(x)$, e.g., smooth with compact support) and try to renormalize by taking a limit in which the support of Λ shrinks to a point; alternatively, one may use interior-boundary conditions, see Chap. 5. Here, we will just ignore this problem and proceed formally. Other problems that we will ignore are that $g \in \mathbb{C}^4$ is not Lorentz invariant, and that there are negative energy solutions.

4.3 Multi-time Emission–Absorption Model

We can now formulate the multi-time equations governing the evolution of ψ on the set of spacelike configurations

$$\mathscr{S}_{xy} = \bigcup_{M,N=0}^{\infty} \left\{ \text{all } x_j, \, y_k \text{ spacelike or equal} \right\}. \tag{4.18}$$

With the notation $(x_1, \ldots, x_M) = x^{4M}$ and $(y_1, \ldots, y_N) = y^{4N}$, the equations read:

$$i \frac{\partial \psi}{\partial x_j^0} (x^{4M}, y^{4N}) = H_{x_j}^{\text{Dirac}} \psi(x^{4M}, y^{4N})$$

$$+ \sqrt{N+1} \sum_{s_{N+1}=1}^{4} g_{s_{N+1}}^* \, \psi_{s_{N+1}} \left(x^{4M}, (y^{4N}, x_j) \right) \tag{4.19a}$$

$$+ \frac{1}{\sqrt{N}} \sum_{k=1}^{N} G_{s_k} (y_k - x_j) \, \psi_{\widehat{s_k}} (x^{4M}, y^{4N} \setminus y_k),$$

$$i \frac{\partial \psi}{\partial y_k^0} (x^{4M}, y^{4N}) = H_{y_k}^{\text{Dirac}} \psi(x^{4M}, y^{4N}) \tag{4.19b}$$

with Green's function G the solution to

$$i \frac{\partial G}{\partial t} = H_y^{\text{Dirac}} G \tag{4.20}$$

with initial condition

$$G_s(t = 0, y) = g_s \, \delta^3(y). \tag{4.21}$$

The second and third line of (4.19a) represent the interaction at collision configurations.

Several remarks about these equations are in order:

- The meaning of (4.19) at, e.g., $x_j = y_k$ (where the x_j^0 direction leaves the domain $\mathscr{S}_{xy}$) is that the directional derivative

$$i\left(\frac{\partial}{\partial x_j^0} + \frac{\partial}{\partial y_k^0}\right)\psi \qquad (4.22)$$

  (which remains in $\mathscr{S}_{xy}$) is given by the sum of (4.19a) and (4.19b).
- If all times are equal, we recover the one-time Hamiltonian of Sect. 4.2.
- $G(y_k - x_j) = 0$ if y_k is spacelike separated from x_j. As a consequence, we could also add $\sum_j G_{s_k}(\ldots)$ to the y_k^0 equation and remove the last line from (4.19a): there would be no difference on $\mathscr{S}_{xy}$ when $x_j \neq y_k$, and when $x_j = y_k$ then we use the directional derivative anyway. Thus, the alternative equation has the same solution on $\mathscr{S}_{xy}$ (but a different solution on $\Gamma(\mathbb{R}^4)^2$).
- The 0-particle sector has no time variable in the multi-time formulation, so we need a one-time theory with time-independent 0-particle sector ("no creation out of the vacuum").
- The evolution (4.19) preserves permutation symmetry for space-time points, i.e.,

$$\psi_{r_{\pi(1)}\ldots r_{\pi(M)} s_{\varrho(1)}\ldots s_{\varrho(N)}}(x_{\pi(1)}, \ldots, x_{\pi(M)}, y_{\varrho(1)}, \ldots, y_{\varrho(N)})$$
$$= (-1)^\pi \, \psi_{r_1\ldots r_M s_1\ldots s_N}(x_1, \ldots, x_M, y_1, \ldots, y_N) \qquad (4.23)$$

  for permutations π, ϱ and $(-1)^\pi$ the sign of π.

We then find the following crucial result. We call it an "Assertion" rather than a "Theorem" since a rigorous proof is not available due to the ultraviolet divergence of the model.

Assertion 4.1 *The model (4.19) is consistent, i.e., ignoring the UV divergence, for every initial condition there exists a unique solution on* $\mathscr{S}_{xy}$.

We note in passing that in the special case when $g^\dagger \beta g = 0$ or the mass m_y vanishes, then solutions exist even on $\Gamma(\mathbb{R}^4)^2$, i.e., also on non-spacelike configurations.

Sketch of proof First, computing the consistency conditions (see also Exercise 4.3 in Sect. 4.7) yields

$$\left[i\frac{\partial}{\partial x_i^0} - H_{x_i}, i\frac{\partial}{\partial x_j^0} - H_{x_j}\right] = \sum_{s=1}^{4} g_s^*\big(G_s(x_i - x_j) - G_s(x_j - x_i)\big)$$
$$= 2i\, m_y\, g^\dagger \beta g\, \Delta(x_i - x_j), \qquad (4.24)$$

where Δ is the so-called Pauli–Jordan function, which vanishes if its argument is spacelike or zero. Second, we need to generalize the notion of propagation locality (discussed for one particle in Sect. 1.2) to a variable particle number. Third and finally, we proceed similarly to the δ-range model of Sect. 2.3: we partition the particles into families of equal time coordinate and proceed by induction over the number of families. $\square$

A rigorous version of Assertion 4.1 for a variant of (4.19) with UV cut-off (i.e., delta distributions are approximated by square-integrable functions) has been proved in [37]. To implement the cut-off, the wave function ψ is defined on the set $\mathscr{S}_\delta$ of δ-spacelike configurations (with $\delta > 0$ the cut-off length) as in (2.42), now for a variable number of particles. It is one of the practical advantages of the multi-time formulation over the Tomonaga-Schwinger formulation (involving a ψ_Σ for every Cauchy surface Σ) that it gets along more easily with a UV cut-off.

Another variant of the Emission–Absorption Model, in which the y-particles are not treated as particles but as a field and considered at a common time t_y, was considered and given a multi-time formulation with $M + 1$ time variables by Dirac, Fock, and Podolsky [9] in 1932. Its consistency was made plausible by Bloch [4] in 1934 by guessing what the consistency condition should be and checking it through a calculation. A full proof of the consistency of the Dirac-Fock-Podolsky model (in a version with UV cut-off) was given only recently by Deckert and Nickel [7]. Another model involving multi-time wave functions on $\cup_{N=0}^{\infty} \mathscr{M}^N$ for a variable number of spinless Klein–Gordon particles of a single species was considered by Droz-Vincent [10, 11].

4.4 Pair Creation Model

We discussed the Emission–Absorption Model in detail to exemplify the validity of the multi-time framework in QFT, and to explicitly see the exact form of the multi-time equations and the consistency condition. Of course, the multi-time framework is not restricted to only two particle species and the interaction coming from Hamiltonians of the form (4.16). Let us briefly consider a model with three species of particles, called x, $\overline{x}$, and y; we have in mind that $\overline{x}$ is a simple model of the anti-particle of x. The model allows for the reaction $x + \overline{x} \to y$ (pair annihilation) and $y \to x + \overline{x}$ (pair creation). The interaction Hamiltonian can be written as

$$H_{\text{int}} = \int \mathrm{d}^3x \sum_{r,\overline{r},s=1}^{4} \left(g_{r\overline{r}s}\, a_r^\dagger(x)\, \overline{a}_{\overline{r}}(x)\, b_s(x) + g_{r\overline{r}s}^*\, a_r(x)\, \overline{a}_{\overline{r}}(x)\, b_s^\dagger(x) \right). \quad (4.25)$$

The same remarks as in Sect. 4.3 apply, and similarly a multi-time version can be formulated [45]. What is interesting to note here is that in the consistency conditions factors of $\varepsilon_x\,\varepsilon_{\overline{x}}\,\varepsilon_y - 1$ appear (recall $\varepsilon = +1$ for bosons and $\varepsilon = -1$ for fermions). Thus, the equations are consistent if and only if either all species $x\overline{x}y$ are bosons or two of them are fermions and one boson; that is, consistency already enforces fermion number conservation.

4.5   Heisenberg Picture

While we have focused our discussion on the evolution of wave functions, i.e., the Schrödinger picture, there is also the Heisenberg picture, in which the operators describing the statistics of experiments evolve in time. In non-relativistic quantum mechanics, the equivalence of the two pictures can be expressed by

$$\langle \psi(t) | A\psi(t) \rangle = \langle e^{-iHt}\psi(0) | A e^{-iHt}\psi(0) \rangle = \langle \psi(0) | \underbrace{e^{iHt} A e^{-iHt}}_{A(t)} \psi(0) \rangle ,   \quad (4.26)$$

where A is some (self-adjoint) operator. Then $A(t)$ is the solution of the evolution equation

$$i\frac{\mathrm{d}A(t)}{\mathrm{d}t} = -[H, A(t)] .   \quad (4.27)$$

In QFT, one correspondingly defines

$$a_s^\sharp(t, \boldsymbol{x}) = e^{iHt}\,a_s^\sharp(0, \boldsymbol{x})\,e^{-iHt} ,   \quad (4.28)$$

where $a^\sharp$ means either $a^\dagger$ or a. The basic relation to multi-time wave functions ψ is that

$$\psi(x_1, \ldots, x_N) = \frac{1}{\sqrt{N!}}\langle \varnothing | a(x_1)\cdots a(x_N)\Psi_0 \rangle ,   \quad (4.29)$$

where $|\varnothing\rangle$ is the Fock vacuum characterized by $a(x)|\varnothing\rangle = 0$, and Ψ_0 the vector in Hilbert space that is fixed in the Heisenberg picture.

Remarks.

- If we set all times equal, then (4.29) reproduces the single-time wave function φ, provided there is no creation out of the vacuum, i.e.,

$$e^{-iHt}|\varnothing\rangle = |\varnothing\rangle .   \quad (4.30)$$

Indeed,

$$\psi(t, \boldsymbol{x}_1, \ldots, t, \boldsymbol{x}_n)$$

$$= \frac{1}{\sqrt{N!}} \langle \emptyset | e^{iHt} a(\boldsymbol{x}_1) e^{-iHt} e^{iHt} a(\boldsymbol{x}_2) e^{-iHt} \cdots e^{iHt} a(\boldsymbol{x}_N) e^{-iHt} \Psi_0 \rangle$$

$$= \frac{1}{\sqrt{N!}} \langle \emptyset | a(\boldsymbol{x}_1) a(\boldsymbol{x}_2) \cdots a(\boldsymbol{x}_N) \varphi_t \rangle$$

$$= \frac{1}{\sqrt{N!}} \langle a^{\dagger}(\boldsymbol{x}_N) \cdots a^{\dagger}(\boldsymbol{x}_1) \emptyset | \varphi_t \rangle$$

$$= \frac{1}{N!} \int d^3 \boldsymbol{y}_1 \cdots \int d^3 \boldsymbol{y}_N \sum_{\pi \in S_N} \varepsilon^{\pi} \times$$

$$\times \delta^3(\boldsymbol{y}_1 - \boldsymbol{x}_{\pi(1)}) \cdots \delta^3(\boldsymbol{y}_N - \boldsymbol{x}_{\pi(N)}) \varphi_t(\boldsymbol{y}_1, \ldots, \boldsymbol{y}_N)$$

$$= \varphi_t(\boldsymbol{x}_1, \ldots, \boldsymbol{x}_N) . \tag{4.31}$$

Here, S_N is the permutation group of $\{1, \ldots, N\}$, and, as usual, $\varepsilon = +1 = \varepsilon^{\pi}$ for bosons, while for fermions, $\varepsilon = -1$ and ε^{π} means the sign of π.

• On collision-free spacelike configurations, i.e., those with $x_i \neq x_j$ for $i \neq j$, (4.29) can be rewritten in the form

$$\psi(x_1, \ldots, x_N) = \frac{1}{\sqrt{N!}} \langle \emptyset | \Phi(x_1) \cdots \Phi(x_N) \Psi_0 \rangle \tag{4.32}$$

with the field operator $\Phi(x) = a(x) + a^{\dagger}(x)$. This can be seen when we start from the right-hand side of (4.32) and write it as

$$\frac{1}{\sqrt{N!}} \langle \emptyset | (a(x_1) + a^{\dagger}(x_1)) \cdots (a(x_N) + a^{\dagger}(x_N)) \Psi_0 \rangle . \tag{4.33}$$

We multiply out the sums, then each term involves a product $a^{\sharp}(x_1) \cdots a^{\sharp}(x_N)$; by the canonical commutation relation (4.8b), the commutator of $a(x_i)$ and $a^{\dagger}(x_j)$ vanishes (for $i \neq j$) at a collision-free spacelike configuration, so we can move all $a^{\dagger}$ factors to the left, and then to the other side of the scalar product turning them into a operators, which vanish on the vacuum vector. The only term that remains is the right-hand side of (4.29).

• The permutation symmetry of ψ can be read off from (4.32) (or (4.29)). If, at a collision-free configuration, we exchange x_i and x_j among the arguments of ψ, then we exchange the factors $\Phi(x_i)$ and $\Phi(x_j)$ in (4.32). Suppose $i < j$. Since the Φ operators also satisfy the canonical (anti-)commutation relations, we can move a factor $\Phi(x_j)$ one step at the expense of a factor ε. To exchange $\Phi(x_i)$ and $\Phi(x_j)$, we can move $\Phi(x_j)$ to the left by $j - i$ steps and then $\Phi(x_i)$ to the right by $j - i - 1$ steps; the prefactors combine to an overall factor of ε, i.e.,

$$\psi(x_j, x_i) = \varepsilon \, \psi(x_i, x_j) , \tag{4.34}$$

where all other arguments are unchanged (and spin indices are exchanged along with the space-time variables). That is, multi-time wave functions have the same permutation symmetry under exchange of space-time points as single-time wave functions under exchange of space points.

For the Emission–Absorption Model from Sect. 4.3, we can formulate the following explicit result [43].

Assertion 4.2 *Let ψ be a solution to the Emission–Absorption Model (4.19) with initial condition ψ_0. Then, on $\mathscr{S}_{xy}$,*

$$\psi_{r_1\ldots r_M s_1\ldots s_N}(x_1,\ldots,x_M,y_1,\ldots,y_N) =$$
$$\frac{(-1)^{M(M-1)/2}}{\sqrt{M!N!}}\langle\emptyset|a_{r_1}(x_1)\cdots a_{r_M}(x_M)b_{s_1}(y_1)\cdots b_{s_N}(y_N)\psi_0\rangle \qquad (4.35)$$

Sketch of proof We use the Tomonaga-Schwinger equation (see Sect. 4.6) to conclude that, for a Cauchy surface Σ and $x \in \Sigma$,

$$a_r(x) = U_{\Sigma}^{\Sigma_0}\, a_{\Sigma,r}(x)\, U_{\Sigma_0}^{\Sigma}, \qquad (4.36)$$

where $U_{\Sigma}^{\Sigma_0}$ is the time evolution from Σ to the surface $\Sigma_0 = \{t = 0\}$, and $a_{\Sigma,r}(x)$ is the annihilation operator on Σ, which satisfies the CCR (or CAR) on Σ with suitably transformed δ distributions. Then we choose x^{4M}, y^{4N} on Σ and (making use of no creation out of vacuum, $U|\emptyset\rangle = |\emptyset_{\Sigma}\rangle$, with $U = U_{\Sigma}^{\Sigma_0}$) obtain that

$$\langle\emptyset|a_{r_1}(x_1)\cdots a_{r_M}(x_M)b_{s_1}(y_1)\cdots b_{s_N}(y_N)\psi_0\rangle$$
$$= \langle\emptyset|U^{-1}UaU^{-1}\cdots UaU^{-1}UbU^{-1}\cdots UbU^{-1}\underbrace{U\psi_0}_{\psi_{\Sigma}}\rangle$$
$$= \langle\emptyset_{\Sigma}|a_{\Sigma}\cdots a_{\Sigma}b_{\Sigma}\cdots b_{\Sigma}\psi_{\Sigma}\rangle$$
$$= \sqrt{M!N!}(-1)^{M(M-1)/2}\psi_{\Sigma}(x^{4M},y^{4N}). \qquad (4.37)$$

□

4.6   Tomonaga-Schwinger Approach

Finally, we relate the multi-time formalism to another approach that makes use of the interaction picture. In non-relativistic quantum mechanics, where $H = H^{\text{free}} + V$ and the single-time wave function $\varphi(t)$ in the Schrödinger picture obeys

$$i\frac{d\varphi(t)}{dt} = H\varphi(t), \qquad (4.38)$$

the interaction picture deals with

$$\widetilde{\varphi}(t) = e^{-iH^{\text{free}}t}\varphi(t), \tag{4.39}$$

which obeys

$$
\begin{aligned}
i\frac{d\widetilde{\varphi}(t)}{dt} &= H^{\text{free}}\widetilde{\varphi}(t) + e^{-iH^{\text{free}}t}H\varphi(t) \\
&= H^{\text{free}}\widetilde{\varphi}(t) + e^{-iH^{\text{free}}t}He^{iH^{\text{free}}t}e^{-iH^{\text{free}}t}\varphi(t) \\
&= e^{-iH^{\text{free}}t}Ve^{iH^{\text{free}}t}\widetilde{\varphi}(t) \\
&=: \widetilde{V}(t)\,\widetilde{\varphi}(t).
\end{aligned} \tag{4.40}
$$

In the Tomonaga-Schwinger approach, one considers a wave function ψ_Σ associated with every Cauchy surface Σ. To work in a fixed Hilbert space, one uses the free evolution $F_\Sigma^{\Sigma'}$ to identify $\mathscr{H}_\Sigma$ with $\mathscr{H}_{\Sigma'}$. Fix, say, $\mathscr{H} = \mathscr{H}_{\Sigma_0}$, then

$$\mathscr{H} \ni \widetilde{\psi}_\Sigma = F_\Sigma^{\Sigma_0}\underbrace{\psi_\Sigma}_{\in\mathscr{H}_\Sigma}. \tag{4.41}$$

The Tomonaga-Schwinger equation says that

$$i(\widetilde{\psi}_{\Sigma'} - \widetilde{\psi}_\Sigma) = \left(\int_\Sigma^{\Sigma'} d^4x\,\mathcal{H}_I(x)\right)\widetilde{\psi}_\Sigma \tag{4.42}$$

for infinitesimally close Cauchy surfaces Σ, Σ', where the interaction Hamiltonian density is given by

$$\mathcal{H}_I(x) = e^{-iH^{\text{free}}x^0}H_{\text{int}}(x)\,e^{iH^{\text{free}}x^0}. \tag{4.43}$$

Recall that, e.g., for the Emission–Absorption Model,

$$H_{\text{int}}(x) = a^\dagger(x)\big(b(x) + b^\dagger(x)\big)a(x). \tag{4.44}$$

For this model, one can show the following [43]:

- The multi-time equations (4.19) define unitary time evolution mappings $U_\Sigma^{\Sigma'}$: $\mathscr{H}_\Sigma \to \mathscr{H}_{\Sigma'}$.
- Any solution ψ to the multi-time equations (4.19) defines

$$\psi_\Sigma(x^{4M}, y^{4N}) := \psi(x^{4M}, y^{4N}) \text{ for } x^{4M}, y^{4N} \subset \Sigma \tag{4.45}$$

and

$$\widetilde{\psi}_\Sigma := F_\Sigma^{\Sigma_0}\psi_\Sigma. \tag{4.46}$$

Then $\widetilde{\psi}_\Sigma$ satisfies the Tomonaga-Schwinger equation.

- The operators $\mathcal{H}_I(x)$ defined by (4.43) and (4.44) satisfy the Tomonaga-Schwinger consistency condition

$$[\mathcal{H}_I(x), \mathcal{H}_I(y)] = 0 \qquad (4.47)$$

for spacelike x, y.

- The other way around, in order for a family of wave functions ψ_Σ to fit together to form a multi-time wave function ψ, we need that ψ_Σ and $\psi_{\Sigma'}$ agree at every configuration that lies on both Σ and Σ',

$$\psi_\Sigma(x^{4M}, y^{4N}) = \psi_{\Sigma'}(x^{4M}, y^{4N}) \quad \forall x^{4M}, y^{4N} \subset \Sigma \cap \Sigma'. \qquad (4.48)$$

This is true for ψ_Σ obtained from solutions $\widetilde{\psi}_\Sigma$ of the Tomonaga-Schwinger equation (4.42) with (4.43) and (4.44), i.e., for the Emission–Absorption Model. The proof [43] uses in particular the Fock space structure, the fact that in the Emission–Absorption Model there is no creation from the vacuum, propagation locality, and that $\mathcal{H}_I$ is a local expression.

4.7 Exercises

Exercise 4.1 (multi-time equations for Φ^4 theory) "Φ^4 theory" is a quantum field theory model in which the Heisenberg field operators $\Phi(x)$ obey the evolution equation

$$(\Box + m^2)\Phi(x) = \Phi^3(x). \qquad (4.49)$$

Use this equation and the expression of multi-time wave functions via field operators,

$$\psi^{(N)}(x_1, \ldots, x_N) = \frac{1}{\sqrt{N!}} \langle \emptyset | \Phi(x_1) \cdots \Phi(x_N) \Psi_0 \rangle, \qquad (4.50)$$

to derive multi-time equations for $\psi^{(N)}$. (These equations should only contain $\psi^{(n)}$ for different values for n, not any field operators.)

Exercise 4.2 (creation and annihilation operators) Let us consider the scalar bosonic creation and annihilation operators defined by

$$(a(\boldsymbol{x})\varphi)(\boldsymbol{x}_1, \ldots, \boldsymbol{x}_N) = \sqrt{N+1}\, \varphi(\boldsymbol{x}_1, \ldots, \boldsymbol{x}_N, \boldsymbol{x}),$$

$$(a^\dagger(\boldsymbol{x})\varphi)(\boldsymbol{x}_1, \ldots, \boldsymbol{x}_N) = \frac{1}{\sqrt{N}} \sum_{j=1}^{N} \delta^3(\boldsymbol{x}_j - \boldsymbol{x})\varphi(\boldsymbol{x}_1, \ldots, \widehat{\boldsymbol{x}}_j, \ldots, \boldsymbol{x}_N), \qquad (4.51)$$

where $\widehat{(\cdot)}$ denotes omission.

Show that for any operator $H : L^2(\mathbb{R}^3) \rightarrow L^2(\mathbb{R}^3)$,

$$\int_{\mathbb{R}^3} d^3x\, a^\dagger(\boldsymbol{x}) H_{\boldsymbol{x}} a(\boldsymbol{x}) \varphi(\boldsymbol{x}_1, \ldots, \boldsymbol{x}_N) = \sum_{j=1}^{N} H_{\boldsymbol{x}_j} \varphi(\boldsymbol{x}_1, \ldots, \boldsymbol{x}_N). \qquad (4.52)$$

Exercise 4.3 (*consistency condition for the Emission–Absorption Model*) Recall the Emission–Absorption Model (now with a single x-particle),

$$i\partial_{x^0}\psi = \left(H_x^0 + H_x^{\text{int}} \right)\psi,$$
$$i\partial_{y_k^0}\psi = H_{y_k}^0 \psi, \qquad (4.53)$$

where $H^0 = -i \sum_{a=1}^{3} \gamma^0 \gamma^a \partial_{x^a} + \gamma^0 m$ is the free Dirac Hamiltonian, and

$$\left(H_x^{\text{int}}\psi \right)(x, y^{4N}) = \sqrt{N+1} \sum_{s_{N+1}=1}^{4} g^*_{s_{N+1}} \psi_{s_{N+1}}\left(x, (y^{4N}, x)\right)$$

$$+ \frac{1}{\sqrt{N}} \sum_{k=1}^{N} G_{s_k}(y_k - x)\, \psi_{\widehat{s_k}}\left(x, y^{4N} \setminus y_k\right), \qquad (4.54)$$

with some function $G : \mathbb{R}^4 \rightarrow \mathbb{C}^4$ and $g \in \mathbb{C}^4$. Here, we have assumed that there is only one x-particle. Carefully compute the consistency condition

$$\left[i\partial_{x^0} - H_x, i\partial_{y_k^0} - H_{y_k} \right] = 0. \qquad (4.55)$$

What is a good choice for G to fulfill the condition?

Chapter 5
Interior-Boundary Conditions
for Multi-time Wave Functions

We now extend the model from Lecture 3 to quantum field theory (QFT). Thereby, we also create a rigorous example of a multi-time QFT. Moreover, we will show that the multi-time approach is compatible with "interior-boundary conditions" (IBCs).

5.1 What is an IBC?

It is a linear condition relating the wave function ψ at a boundary point $q \in \partial\Omega$ of the domain Ω to the wave function at an interior point $p \in \Omega$. Usually, $\psi^{(n+1)}(x_1, \ldots, x, \ldots, x, \ldots, x_n)$ is related to $\psi^{(n)}(x_1, \ldots, x, \ldots, x_n)$.

A program of Tumulka, Teufel, and collaborators [51, 52] concerning the development of IBCs has led to the hope that IBCs could help overcome problems with ultraviolet (UV) divergence in QFT. That is because IBCs allow to formulate non-rigorous terms with Dirac δ functions resulting from creation operators in the Hamiltonian rigorously, similarly as contact interactions do with δ potentials in the Hamiltonian.

5.2 Setting

For simplicity, let us focus on a truncated Fock space with 1- and 2-particle sectors only, on massless Dirac particles in 1+1 space-time dimensions, and on creation and annihilation events of the form that 1 particle can split up in 2 or 2 merge into 1. Thus, the wave function is of the form $\psi = (\psi^{(1)}, \psi^{(2)})$ with

$$\psi^{(1)} : \mathbb{R}^2 \to \mathbb{C}^2, \qquad\qquad x := (t, z) \mapsto \psi^{(1)}(t, z) \qquad (5.1a)$$

$$\psi^{(2)} : \overset{\circ}{\mathscr{S}}_1 \to \mathbb{C}^4, \qquad\qquad (t_1, z_1, t_2, z_2) \mapsto \psi^{(2)}(\cdots) \qquad (5.1b)$$

© The Author(s), under exclusive license to Springer Nature Switzerland AG 2020
M. Lienert et al., *Multi-time Wave Functions*,
SpringerBriefs in Physics, https://doi.org/10.1007/978-3-030-60691-6_5

where $z_1 < z_2$, and $\mathscr{S}_1$ has boundary $\mathscr{C}$. In Hamiltonian form, the multi-time equations read

$$i\partial_t \psi^{(1)}(t, z) = H^{\text{Dirac}} \psi^{(1)}(t, z) - A \psi^{(2)}(t, z, t, z) \qquad \text{on } \mathbb{R}^2 \qquad (5.2a)$$

$$i\partial_{t_k} \psi^{(2)}(t_1, z_1, t_2, z_2) = H_k^{\text{Dirac}} \psi^{(2)}(t_1, z_1, t_2, z_2) \qquad \text{on } \overset{\circ}{\mathscr{S}}_1, \qquad (5.2b)$$

where A is a complex 2×4 matrix to be determined later, and $A\psi^{(2)}$ in (5.2a) is a source term from the 2-particle sector.

The interior-boundary condition reads

$$\psi^{(2)}_{-+}(t, z, t, z) - e^{i\theta} \psi^{(2)}_{+-}(t, z, t, z) = B\psi^{(1)}(t, z), \qquad (5.3)$$

where B is a complex 1×2 matrix (i.e., a row vector) to be determined. Note that for $B = 0$, (5.3) reduces to the boundary condition (3.37c).

5.3 Derivation of the IBC from Local Probability Conservation

Here, probability conservation means that

$$P(\Sigma) := \int_\Sigma d\sigma_\mu \, j^\mu + \int_{(\Sigma \times \Sigma) \cap \overset{\circ}{\mathscr{S}}_1} d\sigma_{1\mu} \, d\sigma_{2\nu} \, j^{\mu\nu} \qquad (5.4)$$

does not depend on the choice of the Cauchy surface Σ. Here,

$$j^\mu := \overline{\psi^{(1)}} \gamma^\mu \psi^{(1)} \qquad \text{is the Dirac current for 1 particle;} \qquad (5.5a)$$

$$j^{\mu\nu} := \overline{\psi^{(2)}} \gamma_1^\mu \gamma_2^\nu \psi^{(2)} \qquad \text{is the Dirac current for 2 particles,} \qquad (5.5b)$$

and we obtain the relations

$$\partial_\mu j^\mu(x) = -2 \, \text{Im}\left(\psi^{(1)\dagger}(x) \, A \, \psi^{(2)}(x, x)\right), \qquad (5.6a)$$

which is minus the gain of probability in the 1-particle sector, and

$$\partial_{1\mu} j^{\mu\nu} = 0 = \partial_{2\nu} j^{\mu\nu} \qquad (5.6b)$$

which means that in the 2-particle sector, probability can only be lost through the boundary $\mathscr{C}$. We want that the gain in sector 1 equals the loss in sector 2 (see Fig. 5.1), which is equivalent to

Fig. 5.1 Illustration of the current balance condition (5.7)

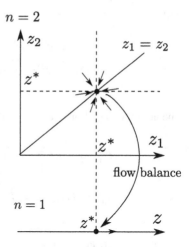

$$2\,\mathrm{Im}\big(\psi^{(1)\dagger}(x)\,A\,\psi^{(2)}(x,x)\big) = (j^{01} - j^{10})(x,x) = 2\big(|\psi^{(2)}_{+-}|^2 - |\psi^{(2)}_{-+}|^2\big)(x,x).$$
(5.7)

(The last equality was derived in (3.35).) This condition needs to be ensured by the IBC. We plug the IBC into (5.7) and find after some calculation:

Theorem 5.1 ([30]) *Eq.* (5.7) *holds if A and B have the form*

$$A = \begin{pmatrix} 0 & \widetilde{A}^{\dagger} & 0 \\ 0 & & 0 \end{pmatrix}$$
(5.8a)

$$\widetilde{A} = \begin{pmatrix} w_1 & w_2 \\ w_1 e^{i\phi} & w_2 e^{i\phi} \end{pmatrix} \text{ with } w_1, w_2 \in \mathbb{C},\ \phi \in [0, 2\pi)$$
(5.8b)

and $B = \frac{1}{2i}(1, e^{i\theta})\widetilde{A}$ *with* $\theta \in [0, 2\pi)$.
(5.8c)

Example 5.1 For $\theta = 0 = \phi$ and $w_1 = 1 = w_2$, one obtains

$$\widetilde{A} = \begin{pmatrix} 1 & 1 \\ 1 & 1 \end{pmatrix} \text{ and } B = -i(1, 1).$$
(5.9)

5.4 Relation to Creation and Annihilation Operators

Considering the model at equal times we obtain a Hamiltonian formulation. The Hamiltonian on Fock space is given by

$$H = H^{\text{free}} + H_{\text{int}}$$
(5.10a)

$$H_{\text{int}} = H^{\text{ann}}_{\text{int}} + H^{\text{cre}}_{\text{int}}$$
(5.10b)

with

$$(H_{\text{int}}^{\text{ann}}\varphi)^{(2)} = 0 \qquad (5.11a)$$

$$(H_{\text{int}}^{\text{cre}}\varphi)_s^{(1)}(t, z) = -\sqrt{2} \sum_{t,u=\pm1} A_s^{tu} \varphi^{(2)}(t, z, z). \qquad (5.11b)$$

The action of $H_{\text{int}}^{\text{ann}}$ on φ agrees with

$$H_{\text{int}}^{\text{ann}} = \int dz \sum_{r,s,t} A_r^{st} a_r^\dagger(z) a_s(z) a_t(z), \qquad (5.12a)$$

$$\text{so } H_{\text{int}}^{\text{cre}} = (H_{\text{int}}^{\text{ann}})^\dagger = \int dz \sum_{r,s,t} (A_r^{st})^* a_t^\dagger(z) a_s^\dagger(z) a_r(z), \qquad (5.12b)$$

where $a^\dagger$ and a are the creation and annihilation operators as defined in Chap. 4.
From the last expression we obtain that

$$(H_{\text{int}}^{\text{cre}}\varphi)_{s_1 s_2}^{(2)}(t, z_1, z_2) = \frac{1}{\sqrt{2}} \sum_r \left[-(A_r^{s_1 s_2})^* + (A_r^{s_2 s_1})^* \right] \delta(z_1 - z_2) \varphi^{(1)}(z_1). \qquad (5.13)$$

The square bracket equals $-2A_r^{s_2 s_1}$ if A is anti-symmetric, as would correspond to
the case $\phi = \pi$. The δ function should be interpreted by integrating over

$$i\partial_t \varphi^{(2)} = H^{\text{free}}\varphi^{(2)} + \underbrace{(H_{\text{int}}^{\text{ann}}\varphi)^{(2)}}_{=0} + (H_{\text{int}}^{\text{cre}}\varphi)^{(2)} \qquad (5.14)$$

in an ε-neighborhood of $z_1 = z_2$ and letting $\varepsilon \to 0$. This yields the IBC for $\phi = \pi = \theta$.

5.5   Sketch of How to Construct the Solution of the Model

Here is the main idea (carried out in detail in [30]). We know:

- Given $\psi^{(2)}$, we can solve the equation for $\psi^{(1)}$,

$$(i\partial_t - H^{\text{Dirac}})\psi^{(1)}(t, z) = -\underbrace{A\psi^{(2)}(t, z, t, z)}_{=:f(t,z)}, \qquad (5.15)$$

uniquely:

$$\psi^{(1)}(t, z) = \begin{pmatrix} \psi_-^{(1)}(0, z - t) \\ \psi_+^{(1)}(0, z + t) \end{pmatrix} + \int_0^t ds \begin{pmatrix} f_-(0, z - t + s) \\ f_+(0, z + t - s) \end{pmatrix}. \qquad (5.16)$$

- Given $\psi^{(1)}$, we can solve the equation for $\psi^{(2)}$ uniquely, using the solution formula (3.38) (or a slight modification thereof).
  Explicitly, let $u_k = z_k - t_k$, $v_k = z_k + t_k$. Then

$$\psi^{(2)}_{--}(t_1, z_1, t_2, z_2) = \psi^{(2)}_{0--}(u_1, u_2) \tag{5.17a}$$

$$\psi^{(2)}_{-+}(t_1, z_1, t_2, z_2) = \begin{cases} \psi^{(2)}_{0-+}(u_1, v_2) & \text{if } u_1 < v_2 \\ e^{i\theta}\psi^{(2)}_{0+-}(v_2, u_1) + B\psi^{(1)}(\frac{v_2-u_1}{2}, \frac{v_2+u_1}{2}) \\ & \text{if } u_1 \geq v_2 \end{cases} \tag{5.17b}$$

$$\psi^{(2)}_{+-}(t_1, z_1, t_2, z_2) = \begin{cases} \psi^{(2)}_{0+-}(v_1, u_2) & \text{if } v_1 < u_2 \\ e^{-i\theta}\left(\psi^{(2)}_{0-+}(u_2, v_1) - B\psi^{(1)}(\frac{v_1-u_2}{2}, \frac{v_1+u_2}{2})\right) \\ & \text{if } v_1 \geq u_2 \end{cases} \tag{5.17c}$$

$$\psi^{(2)}_{++}(t_1, z_1, t_2, z_2) = \psi^{(2)}_{0++}(v_1, v_2). \tag{5.17d}$$

- Set up an iteration scheme based on solving Eqs. (5.16) and (5.17) alternatingly. Prove that it has a unique fixed point in a suitable Banach space. The fixed point is the desired solution.

5.6 Conclusion and Outlook

The model described provides a rigorous example of a multi-time QFT. It is almost fully Lorentz invariant.[1] It demonstrates the compatibility of IBCs and multi-time wave functions and nourishes the hope that IBCs could also work for relativistic QFTs.

The proofs (here and in [30]) were limited to a finite number N of particle sectors; it would be of interest to treat a full Fock space. Furthermore, the model could perhaps be extended to include photons or scalar exchange particles. Could a coupling of the form $(\bar{\psi}\gamma^\mu\psi)(\bar{\psi}\gamma_\mu\psi)$ as in the Thirring model be treated? Finally, it would be desirable to replace 1+1 space-time dimensions by 1+3 dimensions, but it is not clear how such a model can be defined, as the Dirac equation tends not to feel boundaries of too low dimension [22, 49].

---

[1]The use of a fixed matrix A instead of an object whose spin indices transform under Lorentz transformations violates Lorentz invariance. There is reason to hope that this problem does not occur in a model with electrons and photons.

5.7 Exercises

Exercise 5.1 (*probability current balance for an IBC model*) Here is an interior-boundary condition (IBC) for particle creation in a non-relativistic model. Suppose an x-particle is fixed at the origin in $\mathbb{R}^3$, and the number N of y-particles can be 0 or 1, so the y-configuration space is $\mathcal{Q} = \{\emptyset\} \cup \mathbb{R}^3$, and the Hilbert space is $\mathcal{H} = \mathbb{C} \oplus L^2(\mathbb{R}^3)$ (a truncated Fock space). On wave functions $\varphi = (\varphi^{(0)}, \varphi^{(1)}) \in \mathcal{H}$ satisfying the IBC

$$\lim_{y \to 0}\left(|y|\,\varphi^{(1)}(y)\right) = -\frac{mg}{2\pi}\,\varphi^{(0)}, \tag{5.18}$$

where g is a coupling constant, the Hamiltonian is defined by

$$(H\varphi)^{(0)} = \frac{g}{4\pi}\int_{\mathbb{S}^2} d^2\omega \lim_{r \searrow 0} \partial_r\left(r\varphi^{(1)}(r\omega)\right) \tag{5.19}$$

$$(H\varphi)^{(1)}(y) = -\frac{\hbar^2}{2m}\Delta\varphi^{(1)}(y). \tag{5.20}$$

Show that then the following balance condition between the loss of probability in the 1-particle sector at $\mathbf{0} \in \mathbb{R}^3$ and the gain in the 0-particle sector due to the source term on the right-hand side of (5.19) holds:

$$-\lim_{r \searrow 0}\left(r^2 \int_{\mathbb{S}^2} d^2\omega\, \boldsymbol{\omega} \cdot \boldsymbol{j}(r\omega)\right) = \frac{\partial |\varphi^{(0)}|^2}{\partial t}. \tag{5.21}$$

Here, the probability current is defined (as usual) by

$$\boldsymbol{j}(y) = \frac{1}{m}\mathrm{Im}\,\varphi^{(1)}(y)^* \nabla_y \varphi^{(1)}(y). \tag{5.22}$$

Hint: write $y = r\omega$ and note that $\boldsymbol{\omega} \cdot \nabla = \partial_r$.

Chapter 6
Born's Rule for Arbitrary Cauchy Surfaces

Here, we discuss in which sense $|\psi(x_1...x_N)|^2$, with multi-time wave function ψ, yields a detection probability, provided $(x_1...x_N)$ is a spacelike configuration.

6.1 The Curved Born Rule

We formulate a variant of the Born rule for curved surfaces in space-time.

Curved Born rule (CBR). *If detectors are placed along a Cauchy surface Σ, then the probability distribution of the detection events has density $\rho_\Sigma = |\psi_\Sigma|^2$, suitably interpreted.*

"Suitably interpreted" means, for example for a Dirac wave function $\psi_\Sigma(x_1...x_N) \in (\mathbb{C}^4)^{\otimes N}$, that $|\cdot|^2$ is computed using, in each spin space (for x_k), the basis corresponding to the Lorentz frame tangent to Σ at x_k as in Fig. 6.1. Equivalently,

$$|\psi_\Sigma(x_1...x_N)|^2 = \overline{\psi_\Sigma}(x_1...x_N)\, \slashed{n}_1(x_1) \cdots \slashed{n}_N(x_N)\, \psi_\Sigma(x_1...x_N). \qquad (6.1)$$

(The wave function ψ_Σ is just the restriction of the multi-time wave function ψ to Σ^N as in (1.41).)

Along with the Born rule usually comes a collapse rule. For a Cauchy surface, it would read as follows.

Curved collapse rule (CCR). *If detectors on a Cauchy surface Σ found the configuration in $A \subseteq \Sigma^N$, then ψ_Σ collapses to $1_A \psi_\Sigma / \|1_A \psi_\Sigma\|$.*

The usual Born rule, which refers to a measurement of positions at a chosen time, can be thought of as a special case of the curved Born rule with Σ a plane that is horizontal in the chosen Lorentz frame. Let us give it a name, and to the corresponding collapse rule as well:

© The Author(s), under exclusive license to Springer Nature Switzerland AG 2020
M. Lienert et al., *Multi-time Wave Functions*,
SpringerBriefs in Physics, https://doi.org/10.1007/978-3-030-60691-6_6

Fig. 6.1 Lorentz frame
tangent to the Cauchy
surface Σ at x

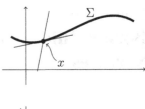

Fig. 6.2 A time t after Σ

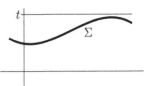

Horizontal Born rule (HBR). *If detectors are placed along $\{x^0 = t\}$, then $\rho_t = |\varphi_t|^2$.*

Horizontal collapse rule (HCR). *If detectors found the configuration in $A \subseteq (\mathbb{R}^3)^N$, then φ_t collapses to $\varphi_{t+} = 1_A \varphi_t / \|1_A \varphi_t\|$.*

Proposition 6.1 *HBR and HCR together determine the statistics of outcomes of any experiment.*

Sketch of proof Any experiment (e.g., a quantum measurement of spin) consists of interaction with some apparatus, which can be represented as a unitary evolution on the Hilbert space of object and apparatus together, followed by reading out the display of the apparatus at some time t after Σ (see Fig. 6.2), which makes use of HBR. For several experiments, one after another, we need to take wave function collapse into account as provided by HCR. (If the second experiment is done on Σ' in the future of Σ, and no horizontal plane fits between Σ and Σ' (as in Fig. 1.11), then we have to use pieces of horizontal planes.)   □

As a consequence, if we assume HBR and HCR, then CBR and CCR are each either false or a theorem. That is, we need a proof of CBR and CCR.

Theorem 6.1 ([34]) *A version of CBR can be proved from HBR and HCR assuming interaction locality and propagation locality.*

The remainder of this chapter will be devoted to discussing this theorem. Its relevance for multi-time wave functions ψ arises when ψ_Σ is obtained from ψ by restriction to Σ^N as in (1.41): Then Theorem 6.1 supports that ψ on $\mathscr{S}$ has direct physical meaning by relating it to the detection probabilities ρ_Σ.

· We already encountered the property of "propagation locality" for the single-particle Dirac equation at the end of Sect. 1.2. We will give a general definition below, as well as one of interaction locality. The "version" of CBR is based of modeling the detection process by approximating Σ through horizontal pieces at times with distance ε as in Fig. 6.3 and taking the limit $\varepsilon \to 0$.

Fig. 6.3 Theorem 6.1 concerns the limit $\varepsilon \to 0$ of the detection probability using detectors placed along horizontal pieces as in the figure

We did not prove CCR because this detection process yields more information than just whether the configuration lies in the set A and thus collapses the wave function more narrowly than to A. It remains for future work to prove the CCR.

Theorem 6.1 is valid for any particle number N, as well as in Fock space (where also superpositions of different particle numbers are allowed), and for several species. Presumably, it can also be extended to curved space-time.

6.2 Some Special Cases

A simple special case, formulated as Corollary 6.1 below, was already proved by Bloch in 1934 [4]; in fact, Bloch proved the following for the non-interacting case:

Theorem 6.2 *Consider N non-interacting particles and $i\partial_{t_k}\psi = H_k\psi$ for $k = 1, \ldots, N$. For each k, choose a time $T_k > 0$ and carry out a position measurement of particle k on $\{x^0 = T_k\}$ (see Fig. 6.4) with result $\boldsymbol{Q}_k$. Assume HBR and HCR. Then*

$$P := \mathrm{Prob}\Big(\boldsymbol{Q}_1 \in d^3\boldsymbol{q}_1, \ldots, \boldsymbol{Q}_N \in d^3\boldsymbol{q}_N\Big)$$
$$= \Big|\psi(T_1, \boldsymbol{q}_1, \ldots, T_N, \boldsymbol{q}_N)\Big|^2 d^3\boldsymbol{q}_1 \cdots d^3\boldsymbol{q}_N. \tag{6.2}$$

Proof Compute P using the single-time wave function φ in $\mathscr{H} = \otimes_k \mathscr{H}_k$, $H = \sum_k H_k$. Without loss of generality, the particles are numbered so that $T_1 \leq T_2 \leq \cdots \leq T_N$. Let P_k be the projection in $\mathscr{H}_k$ to positions in $d^3\boldsymbol{q}_k$. The initial condition is $\varphi(0) = \psi(0...0)$. From HBR we have that

$$\mathrm{Prob}(\boldsymbol{Q}_1 \in d^3\boldsymbol{q}_1) = \| P_1 e^{-iHT_1}\varphi(0)\|^2 . \tag{6.3}$$

Fig. 6.4 Detection of particle k at a chosen time T_k with random result $\boldsymbol{Q}_k$

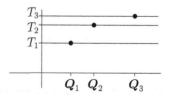

By HCR,

$$\varphi(T_1+) = \frac{P_1 e^{-iHT_1}\varphi(0)}{\|P_1 e^{-iHT_1}\varphi(0)\|}. \tag{6.4}$$

Then, by HBR again,

$$\text{Prob}\left(\boldsymbol{Q}_2 \in d^3\boldsymbol{q}_2 \Big| \boldsymbol{Q}_1 \in d^3\boldsymbol{q}_1\right) = \|P_2 e^{-iH(T_2-T_1)}\varphi(T_1+)\|^2, \tag{6.5}$$

and by HCR,

$$\varphi(T_2+) = P_2 e^{-iH(T_2-T_1)}\varphi(T_1+)/\|\cdots\|, \tag{6.6}$$

and so on, until in the N-th step (and after computing total probabilities from the conditional probabilities) we reach

$$P = \left\| P_N e^{-iH(T_N-T_{N-1})} \cdots P_2 e^{-iH(T_2-T_1)} P_1 e^{-iHT_1}\varphi(0) \right\|^2. \tag{6.7}$$

Using $e^{-iHt} = e^{-iH_1 t} e^{-iH_2 t} \cdots e^{-iH_N t}$ and $[P_k, e^{-iH_j t}] = 0$ for $j \neq k$,

$$P = \|P_N e^{-iH_N(T_N-T_{N-1})} \cdots P_1 e^{-iHT_1}\varphi(0)\|^2 \tag{6.8a}$$

$$= \|P_N e^{-iH_N T_N} \cdots P_1 e^{-iH_1 T_1}\varphi(0)\|^2 \tag{6.8b}$$

$$= |\psi(T_1, \boldsymbol{q}_1, \ldots, T_N, \boldsymbol{q}_N)|^2 \, d^3\boldsymbol{q}_1 \cdots d^3\boldsymbol{q}_N \tag{6.8c}$$

as claimed.   □

Corollary 6.1 *Consider N (possibly interacting) particles in spacelike separated regions $V_k \subset \mathcal{M}$ and detectors along the Cauchy surface Σ. Suppose $\Sigma \cap V_k$ is horizontal for every k (Fig. 6.5). Assume HBR and HCR. Then no interaction occurs between the particles, and by Theorem 6.2, $\rho_\Sigma = |\psi_\Sigma|^2$.*

There are further special cases in which the CBR can be obtained through rather simple arguments. We give two of them.

Bohmian argument for CBR for 1 Dirac particle. At this point, it is useful to assume the perspective of Bohmian mechanics [14, 20]. According to Bohmian mechanics, particles have trajectories which are integral curves of j^μ. Since j^μ is future-causal (i.e., future-timelike or future-lightlike), the integral curves are causal, see Fig. 6.6.

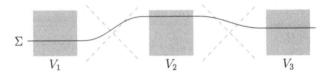

Fig. 6.5 Cauchy surface Σ considered in Corollary 6.1

Fig. 6.6 Example of a future-causal 4-vector field j^μ and its integral curves, the Bohmian trajectories

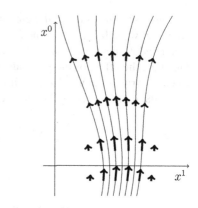

Fig. 6.7 Pieces A_ℓ, Σ_ℓ, and B_ℓ

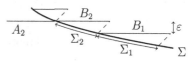

In fact, the trajectory is a random integral curve C with probability distribution given by

$$\text{Prob}\big(C \cap \Sigma \subset \mathrm{d}^3 x\big) = j^\mu(x)\, n_\mu(x)\, \mathrm{d}^3 \sigma(x) \tag{6.9}$$

for every Cauchy surface Σ; $\mathrm{d}^3 \sigma(x)$ means the 3-volume of the 3d set $\mathrm{d}^3 x$ according to the metric as in (1.21). The particle gets detected at $C \cap \Sigma$ (the point where it arrives on Σ), and the presence of detectors on Σ does not influence C before Σ. Hence, the probability of detection in $\mathrm{d}^3 x$ equals the probability that C intersects Σ in $\mathrm{d}^3 x$ in the absence of detectors; the latter quantity is given by $\rho_\Sigma(x)\, \mathrm{d}^3 \sigma(x)$. $\quad\square$

The argument does not work for $N \geq 2$ because Bohmian mechanics is nonlocal: the detection of particle 1 at x_1 can change the trajectory of particle 2 at spacelike separation.

Proof of CBR from HBR and HCR for the 1-particle Dirac equation in 1+1 dimensions. Consider time steps ε, that is, times $\ell\varepsilon$ with $\ell \in \mathbb{Z}$. Let

$$A_\ell = \{x^0 = \ell\varepsilon\} \cap \text{past}(\Sigma)\,, \tag{6.10a}$$

$$\Sigma_\ell = \{x \in \Sigma : (\ell-1)\varepsilon \leq x^0 < \ell\varepsilon\}\,, \tag{6.10b}$$

$$\text{and } B_\ell = \{x^0 = \ell\varepsilon\} \cap \text{future}(\Sigma_\ell) \tag{6.10c}$$

the corresponding horizontal piece as in Fig. 6.7.

At every time step $x^0 = \ell\varepsilon$, we apply a detector in B_ℓ. Let P be the conditional probability of a detection in B_ℓ, given that there was no detection at integer multiples of ε up to $(\ell-1)\varepsilon$; P is given by the integral of $|\widetilde{\psi}|^2$ over B_ℓ, or equivalently by the flux integral of $j^\mu = j_{\widetilde{\psi}}$ over B_ℓ, with $\widetilde{\psi}$ appropriately collapsed at the multiples of ε up to $(\ell-1)\varepsilon$. Now the flux of j through B_ℓ equals the flux of j through Σ_ℓ by the

Ostrogradski–Gauss integral theorem because j is conserved and future-causal. The collapsed wave function at time $(\ell - 1)\varepsilon$ is $\widetilde{\psi} = \mathcal{N}' \, \psi' \, 1_{A_{\ell-1}}$ with $\mathcal{N}'$ a normalizing factor and ψ' the wave function before the last collapse. By propagation locality (see Fig. 1.3), the effect of the previous collapses is included in the factor $1_{A_{\ell-1}}$, so we can also write $\widetilde{\psi} = \mathcal{N} \, \psi \, 1_{A_{\ell-1}}$ with ψ the uncollapsed wave function obtained from the initial data by solving the Dirac equation and

$$\mathcal{N}^{-2} = \int_{A_{\ell-1}} d^3\sigma(x) \, |\psi(x)|^2 \,. \tag{6.11}$$

Thus,

$$P = \mathcal{N}^2 \left(\text{flux of } j_{\psi 1_{A_{\ell-1}}} \text{ through } \Sigma_\ell \right) \tag{6.12a}$$

$$= \frac{\text{flux of } j_\psi \text{ through } \Sigma_\ell}{\text{Prob(no prior detection)}} \,. \tag{6.12b}$$

Since detection events in different B_ℓ's exclude each other, the probability of detection in B_ℓ is given by the flux of j_ψ through Σ_ℓ. This is the statement of the CBR.   □

6.3   Hypersurface Evolution

We want Theorem 6.1 for *every* multi-time evolution. Multi-time equations provide simple formulation of concrete models, but it is hard to say what an *arbitrary* multi-time evolution is. That is why we consider a slightly different concept that is suitable for general, abstract considerations; we formalize it under the name *hypersurface evolution*. It is based on considering a Hilbert space vector ψ_Σ for every Cauchy surface Σ, which is reminiscent of the Tomonaga-Schwinger approach but avoids the interaction picture; its definition consists of the requirements (i)–(vi) below.

(i) For every Cauchy surface Σ, we are given a Hilbert space $\mathscr{H}_\Sigma$.
(ii) For all Σ, Σ', we are given a unitary isomorphism $U_\Sigma^{\Sigma'} : \mathscr{H}_\Sigma \to \mathscr{H}_{\Sigma'}$ ("evolution") such that $U_{\Sigma'}^{\Sigma''} U_\Sigma^{\Sigma'} = U_\Sigma^{\Sigma''}$ and $U_\Sigma^\Sigma = I_\Sigma$. Let

$$\Gamma(\Sigma) = \{ q \subset \Sigma : \#q < \infty \} \tag{6.13}$$

be the space of unordered configurations on Σ (where $\#Q$ means the number of elements of the set Q). A *PVM* (projection-valued measure) on a set S associates with every $A \subseteq S$ a projection $P(A)$ in $\mathscr{H}$ such that $P(A_1 \cup A_2 \cup ...) = P(A_1) + P(A_2) + ...$ if $A_j \cap A_k = \emptyset$ for $j \neq k$ ("σ-additive") and $P(S) = I$. Here are some examples of PVMs:

- On every L^2 space $\mathscr{H} = L^2(S)$ there acts a PVM P on S according to

$$P(A)\psi = 1_A\,\psi\,. \tag{6.14}$$

- The spectral theorem asserts that every self-adjoint operator T is associated with a PVM P on $\mathbb{R}$. For $A \subseteq \mathbb{R}$, $P(A)$ is the projection to (roughly speaking) the span of the generalized eigenvectors with generalized eigenvalues in A.

(iii) For every Σ, we are given a PVM P_Σ on $\Gamma(\Sigma)$ acting on $\mathscr{H}_\Sigma$ ("configuration observable"). Here are some examples of configuration observables P_Σ:

- The bosonic Fock space of $L^2(\Sigma)$ is $L^2(\Gamma(\Sigma))$, and thus automatically equipped with a PVM on $\Gamma(\Sigma)$.
- More generally, both the bosonic and fermionic Fock spaces of $L^2(\Sigma, \mathbb{C}^K)$ are automatically equipped with a PVM on $\Gamma(\Sigma)$.
- As a consequence, the tensor product of two Fock spaces comes with a PVM on $\Gamma(\Sigma) \times \Gamma(\Sigma)$. Now there is a natural mapping $\Gamma(\Sigma) \times \Gamma(\Sigma) \to \Gamma(\Sigma)$: $(q, q') \mapsto q \cup q'$ whose effect amounts to dropping the difference between the two particle species, and this mapping transports the PVM on $\Gamma(\Sigma) \times \Gamma(\Sigma)$ to the desired PVM on $\Gamma(\Sigma)$.

In this formalization, the PVM captures the $|\psi|^2$ distribution in the sense that, on Fock space,

$$\left|\psi_\Sigma(x_1...x_N)\right|^2 d^3\sigma(x_1)\cdots d^3\sigma(x_N) = \left\|P_\Sigma(d^3x_1 \times \cdots \times d^3x_N)\,\psi_\Sigma\right\|^2. \tag{6.15}$$

Note that the N-particle sector of $\mathscr{H}_\Sigma$ is the range of $P(\Gamma_N(\Sigma))$ with

$$\Gamma_N(\Sigma) = \{q \subset \Sigma : \#q = N\} \tag{6.16}$$

the N-particle sector of the configuration space $\Gamma(\Sigma)$.

(iv) P_Σ is absolutely continuous, i.e., $P_\Sigma(A) = 0$ for every set A of volume 0.

(v) Unique vacuum: $\dim(\text{0-particle sector of } \mathscr{H}_\Sigma) = 1$.

(vi) Factorization: If $A, B \subseteq \Sigma$ are disjoint, then $\mathscr{H}_{A\cup B} = \mathscr{H}_A \otimes \mathscr{H}_B$ and

$$P_{A\cup B} = P_A \otimes P_B\,. \tag{6.17}$$

For example, this is true of Fock spaces, both the bosonic Fock space $\mathscr{F}_+$ and the fermionic one $\mathscr{F}_-$:

$$\mathscr{F}_\pm\big(L^2(A \cup B, \mathbb{C}^K)\big) = \mathscr{F}_\pm\big(L^2(A, \mathbb{C}^K)\big) \otimes \mathscr{F}_\pm\big(L^2(B, \mathbb{C}^K)\big), \tag{6.18}$$

a relation analogous to the relation

$$\Gamma(A \cup B) = \Gamma(A) \times \Gamma(B) \tag{6.19}$$

for (unordered) configuration spaces, which is based on identifying the set $q \subset A \cup B$ with the pair $(q \cap A, q \cap B)$. The relation (6.17) is supposed to mean that the projection for a *product* set is a tensor product: If $C \subseteq \Gamma(A)$ and $D \subseteq \Gamma(B)$, so $C \times D \subseteq \Gamma(A \cup B)$ by (6.19), then

$$P_{A \cup B}(C \times D) = P_A(C) \times P_B(D). \tag{6.20}$$

Again, this is automatically true for Fock spaces.

6.4 Interaction Locality

The property of *interaction locality* (**IL**) expresses the absence of interaction terms between spacelike separated regions in the time evolution. For a hypersurface evolution, it can be formalized as follows: *For all Cauchy surfaces* Σ, Σ' *and* $A \subseteq \Sigma \cap \Sigma'$ (Fig. 6.8),

$$U_{\Sigma}^{\Sigma'} = I_A \otimes U_{\Sigma \backslash A}^{\Sigma' \backslash A}, \tag{6.21}$$

where the last factor does not depend on A except through $\Sigma \setminus A$ *and* $\Sigma' \setminus A$.

Since any violation of interaction locality would be expected to allow for superluminal signaling, and since superluminal signaling is believed to be impossible in our universe, it seems that interaction locality is valid in our universe.

Note that interaction locality is not equivalent to Bell locality. The latter means the absence of influences between events in spacelike separated regions. That sounds very similar, but it is not. In fact, Bell's theorem [3, 21] shows that Bell locality is violated, whereas interaction locality apparently holds.

6.5 Propagation Locality

The property of *propagation locality* can be formalized for a hypersurface evolution as follows. For $R \subseteq \Sigma$ define

$$\forall(R) = \{q \in \Gamma(\Sigma) : q \subseteq R\}. \tag{6.22}$$

Definition 6.1 $\psi_{\Sigma} \in \mathscr{H}_{\Sigma}$ is *concentrated in* $A \subseteq \Sigma$ if and only if $\psi_{\Sigma} \in$ range $P_{\Sigma}(\forall(A))$.

Equivalently, the 3-support of ψ_{Σ} (i.e., the set of all $x \in \Sigma$ that occur in any configuration at which ψ_{Σ} does not vanish) is contained in A.

Fig. 6.8 Cauchy surfaces Σ, Σ' with a common part containing A

Fig. 6.9 Grown set of A in Σ'; the shaded set is a part of future$(A) \cup$ past(A)

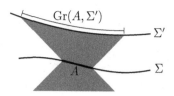

Definition 6.2 For Cauchy surfaces Σ, Σ' and $A \subseteq \Sigma$, the *grown set* of A (Fig. 6.9) is

$$\mathrm{Gr}(A, \Sigma') = \left[\mathrm{future}(A) \cup \mathrm{past}(A) \right] \cap \Sigma'. \tag{6.23}$$

Propagation locality (PL): *Whenever ψ_Σ is concentrated in $A \subseteq \Sigma$, then $\psi_{\Sigma'}$ is concentrated in $\mathrm{Gr}(A, \Sigma')$. Equivalently,*

$$sU_\Sigma^{\Sigma'} \, P_\Sigma(\forall(A)) \, U_{\Sigma'}^{\Sigma} \leq P_{\Sigma'}(\forall(\mathrm{Gr}(A, \Sigma'))). \tag{6.24}$$

Example 6.1 Examples of IL and PL include the following models (always taken for all N so that $\mathscr{H}_\Sigma$ is a Fock space):

- the free multi-time Dirac evolution
- the multi-time Dirac equation in 1+1 dimensions with point interaction as in Chap. 3
- the multi-time Emission–Absorption Model as in Sect. 4.3
- the multi-time Dirac equation in 1+1 dimensions with IBC as in Chap. 5

Upshot. Theorem 6.1 can now be summarized as follows: for hypersurface evolutions,

$$\mathrm{HBR + HCR + IL + PL} \Rightarrow \mathrm{CBR}. \tag{6.25}$$

That is, under reasonable assumptions one can prove that probabilities are given by $|\psi|^2$ at spacelike configurations.

6.6 Exercises

Exercise 6.1 ((PL) $\Rightarrow$ (NCFV)) Explain why for any hypersurface evolution $(\mathscr{H}_\Sigma, P_\Sigma, U_\Sigma^{\Sigma'})_{\Sigma,\Sigma'}$, propagation locality (PL) implies there is no creation from the vacuum (NCFV). The latter property is defined as follows:

$$\mathbf{(NCFV)} \quad U_\Sigma^{\Sigma'} P_\Sigma(\{\emptyset\}) \, U_{\Sigma'}^{\Sigma} = P_{\Sigma'}(\{\emptyset\}). \tag{6.26}$$

(Here, $P_\Sigma(\{\emptyset\})$ is the projection to the vacuum state in $\mathscr{H}_\Sigma$.)

Chapter 7
Multi-time Integral Equations

7.1 Motivation

We have seen in Chap. 2 that differential multi-time equations for a fixed number
N of particles are strongly restricted by the consistency conditions so that physically realistic interactions presumably cannot be implemented in that framework.
Quantum field theories on the other hand, are UV-divergent in the realistic number
of spacetime dimensions (1+3) and hard to make rigorous in a Lorentz invariant
way. (This probably requires a novel idea; however, no such idea is available at
present.) Therefore, we take a different approach in this chapter: direct interactions
(unmediated by fields) with time delay. In classical electrodynamics, Wheeler and
Feynman have shown that using direct interactions instead of fields avoids the UV
problem [62, 63]. The crucial point is that using ψ at non-simultaneous configura-
tions $(t_1, \boldsymbol{x}_1, t_2, \boldsymbol{x}_2)$ with $t_1 \neq t_2$ (in every frame), e.g., the lightlike configuration
$t_2 = t_1 - |\boldsymbol{x}_1 - \boldsymbol{x}_2|$ (see Fig. 7.1), it becomes possible to express interactions with
time delay also on the quantum level. This is a new possibility which is *specific
for multi-time wave functions*! It is not available for the single-time wave function
$\varphi(t, \boldsymbol{x}_1, \boldsymbol{x}_2) = \psi(t, \boldsymbol{x}_1, t, \boldsymbol{x}_2)$. We now show that integral equations are a good way
to define an evolution equation which implements direct interactions for multi-time
wave functions.

7.2 Derivation of a Suitable Integral Equation

Why should the evolution equation be an integral equation? To answer this question,
we follow the derivation in [28]. In fact, an integral equation is a natural possibility
as also the Schrödinger equation

Fig. 7.1 Interaction along
light cones (here between
particle world lines as in the
Wheeler-Feynman theory of
classical electrodynamics)

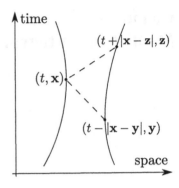

$$(i\partial_t - H_1^{\text{free}} - H_2^{\text{free}} - V)\varphi(t, \boldsymbol{x}_1, \boldsymbol{x}_2) = 0 \qquad (7.1a)$$

$$\varphi(0, \boldsymbol{x}_1, \boldsymbol{x}_2) = \varphi_0(\boldsymbol{x}_1, \boldsymbol{x}_2) \qquad (7.1b)$$

can be reformulated as the integral equation

$$\varphi(t, \boldsymbol{x}_1, \boldsymbol{x}_2) = \varphi^{\text{free}}(t, \boldsymbol{x}_1, \boldsymbol{x}_2) + \int_0^t dt' \int d^3x_1' \, d^3x_2' \times$$

$$G_1^{\text{ret}}(t - t', \boldsymbol{x}_1 - \boldsymbol{x}_1') \, G_2^{\text{ret}}(t - t', \boldsymbol{x}_2 - \boldsymbol{x}_2') \, V(t', \boldsymbol{x}_1', \boldsymbol{x}_2') \, \varphi(t', \boldsymbol{x}_1', \boldsymbol{x}_2') \qquad (7.2)$$

with φ^{free} the solution of (7.1) with $V = 0$ and G_k^{ret} the retarded Green's function of

$$(i\partial_t - H_k^{\text{free}})G^{\text{ret}}(t - t', \boldsymbol{x}_k - \boldsymbol{x}_k') = \delta(t - t') \, \delta^3(\boldsymbol{x}_k - \boldsymbol{x}_k') \,. \qquad (7.3)$$

Now the crucial point is: *Contrary to (7.1), (7.2) has a straightforward relativistic generalization in terms of a multi-time wave function!* Namely:

$$\psi(x_1, x_2) = \psi^{\text{free}}(x_1, x_2) + \int d^4x_1' \, d^4x_2' \, G_1(x_1 - x_1') \, G_2(x_2 - x_2')$$

$$\times K(x_1', x_2') \, \psi(x_1', x_2') \,. \qquad (7.4)$$

Here, ψ^{free} is a solution of free multi-time equations, e.g., Dirac equations

$$D_1\psi^{\text{free}}(x_1, x_2) = 0 \qquad (7.5a)$$

$$D_2\psi^{\text{free}}(x_1, x_2) = 0 \,, \qquad (7.5b)$$

G_k are Green's functions (retarded or otherwise) of these equations,

$$D_k G_k(x_k - x_k') = \delta^4(x_k - x_k') \,, \qquad (7.6)$$

and the function K, the so-called "interaction kernel", is responsible for the interaction, analogous to V in (7.2). Simultaneous interactions correspond to interaction kernels of the form

$$K(t_1, \boldsymbol{x}_1, t_2, \boldsymbol{x}_2) = \delta(t_1 - t_2) V(t_1, \boldsymbol{x}_1, \boldsymbol{x}_2),\qquad(7.7)$$

while interactions along light cones arise from

$$K(x_1, x_2) \propto \delta\left((x_1 - x_2)^2\right)\qquad(7.8)$$

with $(x_1 - x_2)^2 = (x_1^0 - x_2^0)^2 - |\boldsymbol{x}_1 - \boldsymbol{x}_2|^2$.

7.3 How to Understand the Time Evolution

Previously, for differential multi-time equations, the consistency condition was a way to ensure that the many multi-time equations do not contradict one another, and that one obtains a solution starting from arbitrary initial data. Now, we just have a single integral equation, so that a problem of this kind does not seem to occur. However, it is still important to ask: Is there an argument showing that (7.4) has sufficiently many solutions (e.g., for a class of arbitrary initial data)? To answer this question, we note that (7.4) has the following schematic form:

$$\psi = \psi^{\text{free}} + A\psi\qquad(7.9)$$

with $\psi \in \mathscr{B}$, where $\mathscr{B}$ is a suitable Banach space. Now, the important point is that there are several scenarios in which (7.9) makes sense (i.e., defines a time evolution):

(a) If $\|A\| < 1$: then there exists a unique solution for every ψ^{free} by virtue of Banach's fixed point theorem.
(b) If A is a compact operator and $A\psi = 0$ implies $\psi = 0$: then the same result follows by virtue of the Fredholm alternative.

We conclude that in these cases, there are many solutions of (7.9), and the equation indeed works as a time evolution equation. We can read (7.9) as determining a correction to ψ^{free} due to interaction. This is a *scattering picture*: a free incoming solution gets modified as a consequence of the interaction. But contrary to the scattering theory of QFT, we also know the solution for finite times.

7.4   Non-relativistic Limit

Consider $K(x_1, x_2) = \delta\big((x_1 - x_2)^2\big)$. We want to show: A solution ψ of (7.4) approximately solves (7.1) *at equal times* if the time delay of the interaction $|x_1 - x_2|/c$, $c = 1$, is negligible, i.e., if it is allowed to replace

$$\delta\big((t_1 - t_2)^2 - |x_1 - x_2|^2\big) = \frac{1}{2|x_1 - x_2|}\Big[\delta(t_1 - t_2 - |x_1 - x_2|) + \delta(t_1 - t_2 + |x_1 - x_2|)\Big] \tag{7.10}$$

in the integrals over ψ by

$$\frac{1}{|x_1 - x_2|}\delta(t_1 - t_2). \tag{7.11}$$

Consider (7.4) for $t_1 = t_2 = t$:

$$\psi(t, x_1, t, x_2) = \psi^{\text{free}}(t, x_1, t, x_2) + \int dt_1'\, d^3x_1'\, dt_2'\, d^3x_2'\, G_1(t - t_1', x_1 - x_1') \times$$
$$\times\, G_2(t - t_2', x_2 - x_2')\, \underbrace{\delta\big((x_1' - x_2')^2\big)}_{\text{replace as in (7.11)}}\, \psi(t_1', x_1', t_2', x_2') \tag{7.12}$$

$$= \psi^{\text{free}}(t, x_1, t, x_2) + \int dt_1'\, d^3x_1'\, d^3x_2'\, G_1(t - t_1', x_1 - x_1') \times$$
$$\times\, G_2(t - t_1', x_2 - x_2')\, \frac{1}{|x_1' - x_2'|}\, \psi(t_1', x_1', t_1', x_2'), \tag{7.13}$$

which has the form of (7.2) with $V(t, x_1, x_2) = \frac{1}{|x_1 - x_2|}$ and t' rewritten as t_1'. We conclude: (7.4) indeed yields (7.1) and gives a reason why V is the Coulomb potential!

7.5   Action-at-a-Distance Form of the Multi-time Equations

Next, we show that our integral equations have strong similarities with classical action-at-a-distance theories such as Wheeler-Feynman electrodynamics [62, 63].

Let now

$$K(x_1, x_2) = \gamma_1^\mu\, \gamma_{2\mu}\, \delta\big((x_1 - x_2)^2\big) \tag{7.14}$$

and

$$D_k = i\gamma_k^\mu \partial_{k\mu} - m_k \tag{7.15}$$

the free Dirac operator. By acting on (7.4) with D_1, we obtain that

$$D_1\psi(x_1, x_2) = \int d^4x_2'\, G_2(x_2 - x_2')\, \gamma_1^\mu\, \gamma_{2\mu}\, \delta\big((x_1 - x_2')^2\big)\, \psi(x_1, x_2') \tag{7.16}$$

and analogously

$$D_2\psi(x_1, x_2) = \int d^4x_1' \, G_1(x_1 - x_1') \, \gamma_1^\mu \, \gamma_{2\mu} \, \delta\big((x_1' - x_2)^2\big) \, \psi(x_1', x_2). \quad (7.17)$$

These two equations can be summarized as

$$\left[i\gamma_k^\mu(\partial_{k\mu} - i\widehat{A}_{3-k,\mu}(x_k)) - m_k\right]\psi(x_1, x_2) = 0, \quad k = 1, 2 \qquad (7.18)$$

with

$$\widehat{A}_{3-k,\mu}(x_k) \, \psi(x_1, x_2) =$$
$$\int d^4x_{3-k} \, G_{3-k}(x_{3-k} - x_{3-k}') \, \gamma_{3-k,\mu} \, \delta\big((x_k - x_{3-k}')^2\big) \, \psi(x_k, x_{3-k}') \qquad (7.19)$$

where the arguments (x_k, x_{3-k}') are ordered accordingly. We see that $\widehat{A}_1$ acts only on the degrees of freedom associated with particle 2 and $\widehat{A}_2$ only on those associated with particle 1. That is, we have obtained a Dirac equation with a field extracted from the respective other particle's degrees of freedom!

Compare this with the QED model by Dirac from 1932 [8]: Here, $\psi = \psi(x_1, x_2, \mathscr{A})$, where $\mathscr{A}$ represents the field degrees of freedom, obeys the multi-time equations

$$\left[i\gamma_k^\mu(\partial_{k\mu} - i\widehat{A}_\mu(x_k)) - m_k\right]\psi(x_1, x_2, \mathscr{A}) = 0, \quad k = 1, 2, \qquad (7.20)$$

where $\widehat{A}_\mu(x)$ is the electromagnetic field operator. It satisfies

$$\Box\widehat{A}(x) = 0. \qquad (7.21)$$

For Dirac, the principle of interaction of (7.20) was [8, p. 459]:

> "The interaction of the two electrons is due to the motion of each being connected with the same field."

Correspondingly, we can summarize (7.18) by saying:

> "The interaction of the two electrons is due to the motion of each being connected with the field generated by the other."

We conclude that the structure of the equations (7.18) (which are equivalent to our integral equation (7.4)) is indeed similar to that of the classical action-at-a-distance electrodynamics due to Wheeler and Feynman [62, 63].

7.6 Mathematical Analysis of the Integral Equation

Does one of the scenarios mentioned in Sect. 7.3 apply to (7.4)? We analyze this at the example of a simplified class of equations:

1. We consider Klein–Gordon (KG) particles instead of Dirac particles.
2. We focus on the retarded cases $G = G^{\text{ret}}$.
3. We assume a beginning in time (cut off the time integrals before $t = 0$).
4. We consider bounded or only mildly singular interaction kernels K.
5. We focus on the massless case for simplicity.

To explain the motivation, 2 and 3 together lead to time integrals only from 0 to t ("Volterra structure"), and assumption 3 is motivated by the Big Bang singularity. Under these assumptions, we have

$$G_k^{\text{ret}}(t_k, \boldsymbol{x}_k) = \frac{1}{4\pi} \frac{\delta(t_k - |\boldsymbol{x}_k|)}{|\boldsymbol{x}_k|}, \tag{7.22}$$

and (7.4) is equivalent to

$$\psi(x_1, x_2) = \psi^{\text{free}}(x_1, x_2) + \frac{1}{(4\pi)^2} \int_0^\infty dt_1' \int d^3 x_1' \int_0^\infty dt_2' \int d^3 x_2' \times$$

$$\times \frac{\delta(t_1 - t_1' - |\boldsymbol{x}_1 - \boldsymbol{x}_1'|)}{|\boldsymbol{x}_1 - \boldsymbol{x}_1'|} \frac{\delta(t_2 - t_2' - |\boldsymbol{x}_2 - \boldsymbol{x}_2'|)}{|\boldsymbol{x}_2 - \boldsymbol{x}_2'|} (K\psi)(x_1', x_2') \tag{7.23}$$

$$= \psi^{\text{free}}(x_1, x_2) + \frac{1}{(4\pi)^2} \int_{B_{t_1}(\boldsymbol{x}_1)} d^3 x_1' \int_{B_{t_2}(\boldsymbol{x}_2)} d^3 x_2' \times$$

$$\times \frac{1}{|\boldsymbol{x}_1 - \boldsymbol{x}_1'|} \frac{1}{|\boldsymbol{x}_2 - \boldsymbol{x}_2'|} (K\psi)(t_1 - |\boldsymbol{x}_1 - \boldsymbol{x}_1'|, \boldsymbol{x}_1', t_2 - |\boldsymbol{x}_2 - \boldsymbol{x}_2'|, \boldsymbol{x}_2'). \tag{7.24}$$

We can study (7.24) as an equation on the *Banach space*

$$\mathscr{B} := L^\infty\big([0, T]^2, L^2(\mathbb{R}^6)\big) \tag{7.25}$$

with the norm

$$\|\psi\|_{\mathscr{B}} = \operatorname*{ess\,sup}_{t_1, t_2 \in [0,T]} \big\|\psi(t_1, \cdot, t_2, \cdot)\big\|_{L^2(\mathbb{R}^6)}. \tag{7.26}$$

The main result then is the following:

Theorem 7.1 (see theorems 4 and 5 in [35]) *Let $K : \mathbb{R}^8 \to \mathbb{C}$ be bounded or of the form*

$$K(t_1, \boldsymbol{x}_1, t_2, \boldsymbol{x}_2) = \frac{f(t_1, \boldsymbol{x}_1, t_2, \boldsymbol{x}_2)}{|\boldsymbol{x}_1 - \boldsymbol{x}_2|} \tag{7.27}$$

with $f : \mathbb{R}^8 \to \mathbb{C}$ *bounded. Then for every* $\psi^{\text{free}} \in \mathscr{B}$, (7.24) *has a unique solution* $\psi \in \mathscr{B}$.

This result rigorously establishes that multi-time integral equations can indeed be used to define an interacting time evolution for a multi-time wave function!

7.7 Idea of the Proof

Let A be the integral operator in (7.24). The basic idea is to show that the Neumann series

$$\psi = \sum_{k=0}^{\infty} A^k \psi^{\text{free}} \tag{7.28}$$

converges and yields a solution to $\psi = \psi^{\text{free}} + A\psi$. The hard part is to prove the convergence. The strategy for this is to show first that, schematically,

$$\left\| (A\psi)(t_1, \cdot, t_2, \cdot) \right\|_{L^2}^2 \leq \|K\|_\infty \, P(t_1, t_2) \int_0^{t_1} d\rho_1 \int_0^{t_2} d\rho_2 \, \|\psi(\rho_1, \cdot, \rho_2, \cdot)\|_{L^2}^2 , \tag{7.29}$$

where P is a polynomial, and to iterate this relation to obtain that, qualitatively,

$$\left\| (A^k \psi)(t_1, \cdot, t_2, \cdot) \right\|_{L^2}^2 \leq \frac{t_1^k t_2^k \|K\|_\infty^k}{(k!)^2} \|\psi^{\text{free}}\|_{\mathscr{B}}^2 . \tag{7.30}$$

Such a relation yields the absolute convergence of the series (7.28) and proves that A is a bounded operator. Therefore,

$$A \sum_{k=0}^{\infty} A^k \psi^{\text{free}} = \sum_{k=0}^{\infty} A^{k+1} \psi^{\text{free}} = \sum_{k=0}^{\infty} A^k \psi^{\text{free}} - \psi^{\text{free}} \tag{7.31}$$

or

$$A\psi = \psi - \psi^{\text{free}} , \tag{7.32}$$

as desired.

Uniqueness of the solution, given ψ^{free}, follows from $A(\psi_1 - \psi_2) = \psi_1 - \psi_2$ for any two solutions ψ_1, ψ_2, and from the absolute convergence of

$$\sum_{k=1}^{\infty} A^k (\psi_1 - \psi_2) = \sum_{k=0}^{\infty} (\psi_1 - \psi_2) , \tag{7.33}$$

which can only hold for $\psi_1 = \psi_2$.

7.8   On the Cauchy Problem

Let $\psi^{\text{free}} \in \mathscr{B}$ be a solution of the massless Klein-Gordon equations

$$\Box_1 \psi^{\text{free}} = 0 = \Box_2 \psi^{\text{free}} . \tag{7.34}$$

Then ψ^{free} is determined by the following Cauchy data at $t_1 = t_2 = 0$:

$$\psi^{\text{free}}(0, x_1, 0, x_2) = \psi_0^{\text{free}}(x_1, x_2) , \tag{7.35a}$$

$$\partial_{t_1} \psi^{\text{free}}(0, x_1, 0, x_2) = \psi_1^{\text{free}}(x_1, x_2) , \tag{7.35b}$$

$$\partial_{t_2} \psi^{\text{free}}(0, x_1, 0, x_2) = \psi_2^{\text{free}}(x_1, x_2) , \tag{7.35c}$$

$$\partial_{t_1} \partial_{t_2} \psi^{\text{free}}(0, x_1, 0, x_2) = \psi_3^{\text{free}}(x_1, x_2) . \tag{7.35d}$$

As ψ^{free} determines ψ uniquely, these data determine ψ. Moreover, (7.24) implies that

$$\psi(0, x_1, 0, x_2) = \psi^{\text{free}}(0, x_1, 0, x_2) , \tag{7.36}$$

so the data for ψ^{free} are data for ψ as well. (For the first time derivatives this is also the case; however, for the mixed time derivative, $\partial_{t_1} \partial_{t_2} \psi$ may differ from $\partial_{t_1} \partial_{t_2} \psi^{\text{free}}$. This is an artifact from the second-order nature of the Klein-Gordon equation. For the physically more relevant Dirac equation one would not need any data for the time derivatives of ψ and the problem would not occur.)

We conclude that ψ *is uniquely determined by Cauchy data for* ψ^{free} *at the initial time.* However, this is not true at other times as, in general, $\psi^{\text{free}}(t_1, \cdot, t_2, \cdot) \neq \psi(t_1, \cdot, t_2, \cdot)$.

7.9   Conclusions

- The proposed type of multi-time integral equations yield interacting relativistic quantum dynamics in 1+3 dimensions.
- No problem with consistency conditions arises.
- They form a natural generalization of the integral version of the Schrödinger equation.
- They reduce to the Schrödinger equation with Coulomb potential in the non-relativistic limit (neglecting time delay).
- They provide a direct interaction between the charges at the quantum level, and thus a quantum version of action-at-a-distance (Wheeler–Feynman) electrodynamics.
- A rigorous demonstration of the mathematical consistency has been achieved in certain cases.
- The multi-time integral equations feature an unusual role of the time evolution: The interacting wave function is characterized as a correction to a free solution.

7.10 Outlook

- *Curved space-time with Big Bang singularity.* Equations similar to (7.24) have
  been formulated for such space-times [36]; they are particularly appropriate there
  as indeed space-time begins with a certain spacelike surface that we may call
  $t = 0$.
- *Dirac particles.* We have presented here the case based on the Klein-Gordon
  equation, but analogous constructions and results are possible on the basis of the
  Dirac equation [31].
- *N particles.* We have presented here the case of 2 particles. The generalization
  to N particles is also possible but one faces the problem that there are different
  mathematically natural possibilities; see [28] for discussion of several possibilities
  and arguments in favor of one of them.
- *"Super consistency condition."* For multi-time equations with time delay, it is
  not obvious whether there is an analogous condition which takes the place of the
  consistency condition. A certain sufficient condition, called the super consistency
  condition, is discussed in [28]. It is automatically satisfied for dynamics coming
  from multi-time integral equations.
- *Singular interactions.* The physically most interesting interaction kernel $\delta\big((x_1 - x_2)^2\big)$ has recently been treated in [32]. This works shows that multi-time integral
  equations work well to formulate a consistent interacting dynamics even for very
  singular interactions.
- *Open challenges.* It would be useful to obtain conservation laws for multi-time
  integral equations, in particular to find an expression that can be interpreted as
  conserved probability. Moreover, it is open how to combine this approach with
  particle creation and annihilation, be it with the Dirac sea or a Fock space.

7.11 Exercises

Exercise 7.1 (derivation of a potential from a multi-time integral equation) Consider
the following multi-time integral equation in 1+1 dimensions:

$$\psi(t_1, z_1, t_2, z_2) = \psi^{\text{free}}(t_1, z_1, t_2, z_2) + \int_{\mathbb{R}^4} dt'_1 \, dz'_1 \, dt'_2 \, dz'_2 \, G_1(t_1 - t'_1, z_1 - z'_1)$$
$$\times \, G_2(t_2 - t'_2, z_2 - z'_2) H(-(t'_1 - t'_2)^2 + |z'_1 - z'_2|^2)\psi(t'_1, z'_1, t'_2, z'_2),$$
$$(7.37)$$

where H denotes the Heaviside function. Show that if one treats the product $G_1(t_1 - t'_1, z_1 - z'_1)G_2(t_2 - t'_2, z_2 - z'_2)\psi(t'_1, z'_1, t'_2, z'_2)$ as constant in the time interval $t'_2 \in [t'_1 - |z'_1 - z'_2|, t'_1 + |z'_1 - z'_2|]$, then the single-time wave function $\varphi(t, z_1, z_2) = \psi(t, z_1, t, z_2)$ satisfies the integral version of the Schrödinger equation with potential
$V(t, z_1, z_2) \propto |z_1 - z_2|$.

Exercise 7.2 (solution of a Volterra integral equation) Let $T > 0$. Show that for $K \in C([0, T]^2)$ the Volterra integral equation

$$f(t) = f_0(t) + \int_0^t dt'\, K(t, t') f(t') \tag{7.38}$$

has a unique solution $f(t) \in C([0, T])$ for every $f_0 \in C([0, T])$.

Hint: Let $\widehat{K}$ be the integral operator defined by $\widehat{K} f = \int_0^t dt'\, K(t, t') f(t')$. Show through a direct computation that $f = \sum_{n=0}^{\infty} \widehat{K}^n f_0$ converges in the Banach space $\mathscr{B} = C([0, T])$ equipped with the supremum norm and solves the equation.

Conclusion

Multi-time wave functions form an active field of research, as is visible from the numerous recent papers in the bibliography. The lectures in the present book provide a glimpse of the present state of the art; they explain why the multi-time approach is incompatible with interaction potentials and show how interaction can be implemented nevertheless by means of particle creation or integral equations. What we personally find particularly intriguing is the simplicity and naturalness of the multi-time equations. When scientists are searching for the fundamental laws of nature, they regard, among the empirically adequate candidates, those equations as particularly promising or convincing that strike them as particularly simple and natural. That is why we are inclined to believe that multi-time equations are a suitable form for the fundamental laws governing the quantum state.

© The Author(s), under exclusive license to Springer Nature Switzerland AG 2020
M. Lienert et al., *Multi-time Wave Functions*,
SpringerBriefs in Physics, https://doi.org/10.1007/978-3-030-60691-6

References

1. Y. Aharonov, D.Z. Albert, Can we make sense out of the measurement process in relativistic quantum mechanics? Phys. Rev. D **24**, 359–371 (1981)
2. D.J. Bedingham, D. Dürr, G.C. Ghirardi, S. Goldstein, R. Tumulka, N. Zanghì, Matter density and relativistic models of wave function collapse. J. Stat. Phys. **154**, 623–631 (2014). http://arxiv.org/abs/1111.1425
3. J.S. Bell, On the Einstein-Podolsky-Rosen paradox. Physics **1**, 195–200 (1964). Reprinted as Chap. 2 in J.S. Bell, *Speakable and Unspeakable in Quantum Mechanics* (Cambridge University Press, Cambridge, 1987)
4. F. Bloch, Die physikalische Bedeutung mehrerer Zeiten in der Quantenelektrodynamik. Physikalische Zeitschrift der Sowjetunion **5**, 301–305 (1934)
5. H.W. Crater, P. Van Alstine, Two-body Dirac equations. Ann. Phys. **148**, 57–94 (1983)
6. D.-A. Deckert, L. Nickel, Consistency of multi-time Dirac equations with general interaction potentials. J. Math. Phys. **57**, 072301 (2016). http://arxiv.org/abs/1603.02538
7. D.-A. Deckert, L. Nickel, Multi-time dynamics of the Dirac-Fock-Podolsky model of QED. *J. Math. Phys.* **60**: 072301 (2019) http://arxiv.org/abs/1903.10362
8. P.A.M. Dirac, Relativistic quantum mechanics. Proc. R. Soc. Lond. A **136**, 453–464 (1932)
9. P.A.M. Dirac, V.A. Fock, B. Podolsky, On quantum electrodynamics. Physikalische Zeitschrift der Sowjetunion 2(6), 468–479, *Reprinted in J* (Dover, Schwinger, Selected Papers on Quantum Electrodynamics (New York, 1932), p. 1958
10. P. Droz-Vincent, Second quantization of directly interacting particles, in *Relativistic Action at a Distance: Classical and Quantum Aspects*, ed. by J. Llosa (Springer, Berlin, 1982), pp. 81–101
11. P. Droz-Vincent, Relativistic quantum mechanics with non conserved number of particles. J. Geom. Phys. **2**(1), 101–119 (1985)
12. D. Dürr, S. Goldstein, K. Münch-Berndl, N. Zanghì, Hypersurface Bohm–Dirac models. Phys. Rev. A **60**, 2729–2736 (1999). http://arxiv.org/abs/quant-ph/9801070
13. D. Dürr, S. Teufel, On the exit statistics theorem of many particle quantum scattering, in *Multiscale Methods in Quantum Mechanics*, ed. by P. Blanchard, G. Dell'Antonio (Birkhäuser, 2004)
14. D. Dürr, S. Teufel, *Bohmian Mechanics* (Springer, Heidelberg, 2009)
15. A.S. Eddington, The charge of an electron. Proc. R. Soc. A **122**(789), 358–369 (1929)
16. J.A. Gaunt, The triplets of helium. Proc. R. Soc. A **122**(790), 513–532 (1929)
17. G.C. Ghirardi, Collapse theories, in *Stanford Encyclopedia of Philosophy*, ed. by E.N. Zalta (2007). published online by Stanford University. http://plato.stanford.edu/entries/qm-collapse/
18. C. Ghiu, C. Udrişte, Multitime controlled linear PDE systems, in *Contemporary Topics in Mathematics and Statistics with Applications*, vol. 1, ed. by A. Adhikari, M.R. Adhikari, Y.P. Chaubey (Asian Books, New Delhi, 2013), pp. 82–109. http://arxiv.org/abs/1201.0256

© The Author(s), under exclusive license to Springer Nature Switzerland AG 2020
M. Lienert et al., *Multi-time Wave Functions*,
SpringerBriefs in Physics, https://doi.org/10.1007/978-3-030-60691-6

19. S. Goldstein, Quantum Theory Without Observers. Physics Today, Part One: March 1998, 42–46. Part Two: April 1998, 38–42
20. S. Goldstein, Bohmian mechanics, in *Stanford Encyclopedia of Philosophy*, ed. by E.N. Zalta (2001). published online by Stanford University. http://plato.stanford.edu/entries/qm-bohm/
21. S. Goldstein, T. Norsen, D.V. Tausk, N. Zanghì, Bell's theorem. Scholarpedia **6**(10), 8378 (2011). http://www.scholarpedia.org/article/Bell%27s_theorem
22. J. Henheik, R. Tumulka, Interior-Boundary Conditions for the Dirac Equation at Point Sources in 3 Dimensions. Preprint (2020). http://arxiv.org/abs/2006.16755
23. M.K.-H. Kiessling, M. Lienert, A.S. Tahvildar-Zadeh, A Lorentz-Covariant Interacting Electron-Photon System in One Space Dimension. Lett. Math. Phys. online first (2020). http://arxiv.org/abs/1906.03632
24. L. Landau, R. Peierls, Quantenelektrodynamik im Konfigurationsraum. Zeitschrift für Physik **62**, 188–200 (1930). English translation: Quantum electrodynamics in configuration space. in *Selected Scientific Papers of Sir Rudolf Peierls With Commentary* ed by R.H. Dalitz, R. Peierls (Singapore, World Scientific, 1997), pp. 71–82
25. M. Lienert, A relativistically interacting exactly solvable multi-time model for two mass-less Dirac particles in 1+1 dimensions. J. Math. Phys. **56**, 042301 (2015). http://arxiv.org/abs/1411.2833
26. M. Lienert, On the question of current conservation for the Two-Body Dirac equations of constraint theory. J. Phys. A Math. Theor. **48**, 325302 (2015). http://arxiv.org/abs/1501.07027
27. M. Lienert, *Lorentz invariant quantum dynamics in the multi-time formalism*. Ph.D. thesis, Mathematics Institute (Ludwig-Maximilians University, Munich, Germany 2015)
28. M. Lienert, Direct interaction along light cones at the quantum level. J. Phys. A Math. Theor. **51**, 435302 (2018). http://arxiv.org/abs/1801.00060
29. M. Lienert, L. Nickel, A simple explicitly solvable interacting relativistic *N*-particle model. J. Phys. A Math. Theor. **48**, 325301 (2015). http://arxiv.org/abs/1502.00917
30. M. Lienert, L. Nickel, Multi-time formulation of creation and annihilation of particles via interior-boundary conditions. Rev. Math. Phys. **32**, 2050004 (2020). http://arxiv.org/abs/1808.04192
31. M. Lienert, M. Nöth, Existence of relativistic dynamics for two directly interacting Dirac particles in 1+3 dimensions. Preprint (2019) http://arxiv.org/abs/1903.06020
32. M. Lienert, M. Nöth, Singular light cone interactions of scalar particles in 1+3 dimensions. Preprint (2020) http://arxiv.org/abs/2003.08677
33. M. Lienert, S. Petrat, R. Tumulka, Multi-time wave functions versus multiple timelike dimensions. Found. Phys. **47**, 1582–1590 (2017). http://arxiv.org/abs/1708.03376
34. M. Lienert, R. Tumulka, Born's rule for arbitrary Cauchy surfaces. Lett. Math. Phys. **110**, 753–804 (2020). http://arxiv.org/abs/1706.07074
35. M. Lienert, R. Tumulka: A new class of Volterra-type integral equations from relativistic quantum physics. J. Integ. Equ. Appl. **31**, 535–569 (2019). http://arxiv.org/abs/1803.08792
36. M. Lienert, R. Tumulka, Interacting relativistic quantum dynamics of two particles on space-times with a Big Bang singularity. J. Math. Phys. **60**, 042302 (2019). http://arxiv.org/abs/1805.06348
37. S. Lill, L. Nickel, R. Tumulka, Consistency Proof for Multi-Time Schrödinger Equations with Particle Creation and Ultraviolet Cut-Off. Preprint (2020). http://arxiv.org/abs/2001.05920
38. C.W. Misner, K.S. Thorne, J.A. Wheeler, *Gravitation* (Princeton University Press, 1973)
39. N.F. Mott, On the interpretation of the relativity wave equation for two electrons. Proceedings of the Royal Society A **124**(794), 422–425 (1929)
40. E. Nelson, Interaction of Nonrelativistic Particles with a Quantized Scalar Field. J. Math. Phys. **5**, 1190–1197 (1964)
41. L. Nickel, *On the Dynamics of Multi-Time Systems*. Ph.D. thesis, Mathematics Institute (Ludwig-Maximilians University, Munich, Germany, 2019)
42. S. Petrat, R. Tumulka, Multi-time Schrödinger equations cannot contain interaction potentials. J. Math. Phys. **55**, 032302 (2014). http://arxiv.org/abs/1308.1065

43. S. Petrat, R. Tumulka, Multi-time wave functions for quantum field theory. Ann. Phys. **345**, 17–54 (2014). http://arxiv.org/abs/1309.0802

44. S. Petrat, R. Tumulka, Multi-time equations, classical and quantum. Proc. R. Soc. A **470**(2164), 20130632 (2014). http://arxiv.org/abs/1309.1103

45. S. Petrat, R. Tumulka, Multi-time formulation of pair creation. J. Phys. A Math. Theor. **47**, 112001 (2014). https://arxiv.org/abs/1401.6093

46. S. Schweber, *An Introduction To Relativistic Quantum Field Theory* (Row, Peterson and Company, 1961)

47. J. Schwinger, Quantum electrodynamics. I. A covariant formulation. Phys. Rev. **74**(10), 1439–1461 (1948)

48. N.A. Sinitsyn, E.A. Yuzbashyan, V.Y. Chernyak, A. Patra, C. Sun, Integrable time-dependent quantum Hamiltonians. Phys. Rev. Lett. **120**, 190402 (2018). http://arxiv.org/abs/1711.09945

49. E.C. Svendsen, The effect of submanifolds upon essential self-adjointness and deficiency indices. J. Math. Anal. Appl. **80**, 551–565 (1981)

50. S. Teufel, Effective N-body dynamics for the massless Nelson model and adiabatic decoupling without spectral gap. Ann. Henri Poincaré **3**, 939–965 (2002). https://arxiv.org/abs/math-ph/0203046

51. S. Teufel, R. Tumulka, Hamiltonians without ultraviolet divergence for quantum field theories. *Quantum Studies: Mathematics and Foundations* online first (2020). http://arxiv.org/abs/1505.04847

52. S. Teufel, R. Tumulka, Avoiding Ultraviolet Divergence by Means of Interior-Boundary Conditions, in *Quantum Mathematical Physics—A Bridge between Mathematics and Physics* ed by F. Finster, J. Kleiner, C. Röken, J. Tolksdorf (Basel, Birkhäuser, 2016), pp. 293–311. http://arxiv.org/abs/1506.00497

53. B. Thaller, *The Dirac Equation* (Springer, 1991)

54. S. Tomonaga, On a relativistically invariant formulation of the quantum theory of wave fields. Progress Theor. Phys. **1**(2), 27–42 (1946)

55. R. Tumulka, A relativistic version of the Ghirardi–Rimini–Weber model. J. Stat. Phys. **125**, 821–840 (2006). http://arxiv.org/abs/quant-ph/0406094

56. R. Tumulka, The 'Unromantic pictures' of quantum theory. J. Phys. A Math. Theor. **40**, 3245–3273 (2007). http://arxiv.org/abs/quant-ph/0607124

57. R. Tumulka, Distribution of the Time at Which an Ideal Detector Clicks. Preprint (2016). http://arxiv.org/abs/1601.03715

58. R. Tumulka, Detection Time Distribution for the Dirac Equation. Preprint (2016). http://arxiv.org/abs/1601.04571

59. R. Tumulka, A relativistic GRW flash process with interaction, in *Do wave functions jump?* ed by V. Allori, A. Bassi, D. Dürr, N. Zanghì (Springer, Berlin, 2020). http://arxiv.org/abs/2002.00482

60. P. Van Alstine, H.W. Crater, A tale of three equations: Breit, Eddington-Gaunt, and two-body Dirac. Found. Phys. **27**(1), 67–79 (1997)

61. R. Werner, Arrival time observables in quantum mechanics. Annales de l'Institute Henri Poincaré, section A **47**, 429–449 (1987)

62. J.A. Wheeler, R.P. Feynman, Interaction with the absorber as the mechanism of radiation. Rev. Modern Phys. **17**, 157–181 (1945)

63. J.A. Wheeler, R.P. Feynman, Classical electrodynamics in terms of direct interparticle action. Rev. Modern Phys. **21**, 425–433 (1949)

Index

© The Author(s), under exclusive license to Springer Nature Switzerland AG 2020
M. Lienert et al., *Multi-time Wave Functions*,
SpringerBriefs in Physics, https://doi.org/10.1007/978-3-030-60691-6

Printed in the United States
By Bookmasters